Dedication

To Jerry and Hilda Richards
Who taught me to look at things, turn them over and look again. Nothing
is truly as it seems in this life. Your observations, road trips and general
patience for this wild child made living my youth in Sunnyvale's suburbia a
thoughtfully interesting period… a span of time that fueled my imagination
enough to make the tall task of writing this history a pure adventure.
Thank you for all of your loving gifts of wonder,
Sally

FIRST EDITION
COPYRIGHT©2000
BY HERITAGE MEDIA CORPORATION
ALL RIGHTS RESERVED. NO PART OF THIS BOOK
MAY BE REPRODUCED IN ANY FORM OR BY
ANY MEANS, ELECTRONIC OR MECHANICAL, INCLUDING
PHOTOCOPYING, WITHOUT PERMISSION
IN WRITING FROM THE PUBLISHER.
ALL INQUIRIES SHOULD BE ADDRESSED TO HERITAGE MEDIA CORP.
ISBN: 1-886483-45-0
LIBRARY OF CONGRESS CARD CATALOG NUMBER: 00-103658
PUBLISHER: C.E. PARKS
EDITOR-IN-CHIEF: LORI M. PARKS
AUTHOR: SALLY RICHARDS
VP/NATIONAL SALES MANAGER: JILL DELEARY
VP/CORPORATE DEVELOPMENT: BART BARICA
CFO: RANDALL PETERSON
PRODUCTION MANAGER: DEBORAH SHERWOOD
MANAGING EDITOR: BETSY BAXTER BLONDIN
ART DIRECTOR: GINA MANCINI
ASSISTANT ART DIRECTOR: SUSIE PASSONS
PROJECT EDITOR: SARA RUFNER
PRODUCTION STAFF: ASTRIT BUSHI, JEFF CATON, DAVE HERMSTEAD, JAY KENNEDY,
VINCENT KORNEGAY, JOHN LEYVA, MARIANNE MACKIE, BOB MINTON,
GAVIN RATTMANN, CHARLIE SILVIA
COORDINATING EDITORS: RENEE KIM, BETSY LELJA,
ELIZABETH LEX, MARY ANN STABILE, ADRIANE WESSELS, JOHN WOODWARD
PROFILE WRITERS: MARY JEANNE AINSLEY, RANDALL FROST, ALLEN GARDINER,
NORA HORN, STACEY KNAPP, MARK LAPIN, GARY PETERSON,
ROBERT J. SERATA, JOHN STICKLER, KIT PALONE STURMAN
HUMAN RESOURCES MANAGER: ELLEN RUBY
ADMINISTRATION: JUAN DIAZ, AZALEA MAES, MAJKA PENNER,
SCOTT REID, PATRICK RUCKER, CORY SOTTEK

PUBLISHED BY

HERITAGE MEDIA CORP.
1954 KELLOGG AVENUE
CARLSBAD, CALIFORNIA 92008
WWW.HERITAGEMEDIA.COM

PRINTED BY HERITAGE MEDIA CORP. IN THE UNITED STATES OF AMERICA
PUBLISHED IN COOPERATION WITH HISTORY SAN JOSÉ
WWW.HISTORYSANJOSE.ORG

Contents

Preface	4
Acknowledgments	6
Introduction	9
Chapter one *Dawn*	10
Chapter two *Midmorning*	28
Chapter three *Noon*	48
Chapter four *Afternoon*	88
Chapter five *Twilight*	108
Partners in Silicon Valley	142
Bibliography	222
Index	224

PREFACE

How do you tell a story from the Valley's beginning and also give a glimpse at its future as told by the visionaries at the helm? Finding 50,000-plus words to describe such a vast amount of time was not easy. Knowing where to begin this book was quite a task, knowing where to end it was even more difficult.

Sometimes, when the words weighed too heavily and it was too beautiful to stay inside, I'd gather my camera equipment, buy some fresh rolls of film and just drive. I put 5,000 plus miles on my car during this project and took nearly 1,000 slides of the Valley doing her thing. I absolutely loved shooting the early morning sunrises and late twilights; at those times it seemed like there was no one else around and I had the still, quiet Valley all to myself. Sometimes the photos would inspire a new story.

Time and space were big issues on this project. How do you learn so much in so little time and condense it into a comprehensible (and hopefully entertaining) format? Like the sand that forms silicon wafers, time slipped by so easily; it seemed only fitting that I write the book divided by the times of the day. Relatively speaking, the period I wrote about seems to have taken place all in the course of one very, very long, expansive day.

Many of the kids in my generation who grew up in the Valley sacrificed their fathers and mothers to the high-tech industry. We grew up latchkey kids and our parents reminded us to eat breakfast on their way out the door. My dad retired from the Navy in 1976 after serving 20 years and three tours in Vietnam and worked at ESL (now TRW), my mom worked at Ampex and other high-tech companies, and both worked in technical positions until their recent retirements. Growing up in the Valley was such an adventure for a kid surrounded by high-tech. We had all of the accoutrements of modern life in my home as a child — the latest hi-fi equipment, CB radios, a huge TV, a microwave, a home computer (and dad always building more; mom looking over his shoulder, inspecting his work for cold soldering joints). I grew up in a house of gadgets — including the video game Pong. Let me tell you, it was quite a thrill to meet Nolan Bushnell in the flesh!

I lived about two miles from Apple's and Hewlett-Packard's Cupertino offices. While writing this book, I went out to do a photo shoot of the orchard next to HP, and found I had gotten there just in time to see a firewood sign leaning against the cyclone fence. The trees had been pulled out by their roots and lay on their sides like a battlefield of dead soldiers; the war lost being progress. I had grown up with that orchard and as many kids did, I plucked cherries off of those trees and popped them right in my mouth, fully enjoying my ill-gotten booty. So, I did the only thing I could to pay back that once beautifully sweet-tasting orchard, I managed to gently dig up a five-foot cherry tree sprout which is now doing quite well and will carry on its legacy 40 miles from its original home.

Time moves on and either you move with it, or you are left behind like those trees in the orchard.

Who would have thought this place I call home would blossom into an area where its technology drives the rest of the world? This is the place where kingdoms are founded in garages and the plans for dynasties are mapped out on napkins in Cafe Borrone. Absolutely anything can happen to you in Silicon Valley — anything. If you have a dream, chances are you can get the venture capital to fund it.

It's a typically gorgeous day on the Peninsula. As I write this on my G-3 laptop, I'm sitting on a bench overlooking Crystal Springs Reservoir. It's beautiful — Kodachrome blue sky, deep blue water and just a hint of a breeze. It's warm and the screen on my laptop is a little difficult to read in this sunlight, but I'll persevere.

My back is to the people who are running, Rollerblading and walking behind me. I'm picking up bits of their conversations, some of the passersby's talk to themselves — actually they're talking into their cell phones, but I only hear one side of things. Some talk about relationships — or lack thereof, of switching platforms, coding issues, stock — I hear a lot about stock. I'm also taking note of the many languages I'm hearing. This place has become the melting pot — give us your engineers, your coders, your IT geniuses. Smart people looking to attain a piece of the American Dream in this place of mythical proportions. A place reinvented many times but newly, relatively speaking, founded on an industry created from one of the most common elements in the world, sand. Who would have thought?

People come here to set trends, develop new technologies and grab a huge piece of the IPO pie. They are haunted, taunted, tortured by visions of success. Dreams of finding riches, of becoming whole. This has always been the motive for those arriving in California, most feel they are missing something in their lives and that California — Silicon Valley — can provide it. She's like the child prodigy always expected to perform at a higher level than other children. Would you expect the same performance of other areas in the country? Silicon Valley is the prodigal daughter, the one striving selflessly to satisfy expectations, not only because she can, but also because she can do so effortlessly.

Now, we enter this next millennium looking for solutions to the problems our growth is causing, but we know we'll be able to solve them. If we've learned one thing from our history, it's that there is nothing a dreamer can't conceive and an engineer can't build. Who would have ever believed this vast amount of fertile land would have given birth to vacuum tubes and silicon wafers? *Sand Dreams & Silicon Orchards,* where will we go from here?

Sally Richards
January 1, 2000
Crystal Springs Reservoir • San Mateo, CA
www.SallyRichards.com

Acknowledgments

First and foremost, I'd like to thank John Waters who recognized this was a project I would revel in — you are fabulous! I especially want to thank Betsy Blondin, my fabulous editor — you are awesome! Thank you, Sara Rufner, project editor, for your help beyond the call of duty.

I'd also like to thank Jeanne Thivierge (you're really way too cool for words) at the Redwood City Library History Room; you shared your archives unselfishly and without expectation. Margaret Kimball at the Department of Special Collections and University Archives at Stanford University, your patience and knowledge are greatly appreciated. Lisa Christiansen and Kathleen Peregrin at the California History Center are incredibly helpful and leant much to my inner database about the Valley of the Heart's Delight. And a grateful thanks to all of the people at the Moffett Field History Museum — you are wonderful! Thank you Tracey Mazur at the Intel archives — you were extremely generous with your time and slides. A big thanks to Jeremy Miller, U.S. director of public relations at TBWA/Chiat/Day, Los Angeles, for your 11th-hour help.

A super-fantabulous thank you to Paul Mortfield who taught me a great deal about astronomy and helped me work through the chapter titles — not to mention took me to Stanford's microwave dish to shoot the awesome photos for Twilight. You are always a source of humor, support and crazy inspiration.

A big thank-you to Mark Hylkema who saved me from further endless hours of due diligence on the ancient histories of the Bay Area. I had been looking for answers about the Bay Area's early history in all the wrong places and probably would have continued spinning my wheels had it not been for our serendipitous meeting.

And thanks to all the people who gave their knowledge freely to me without expectations or dictation: Dane Andrew, Carol Bartz, Carol Beddo, Nolan Bushnell, Christine Comaford, Dr. Carolyn Dean, Harry Farrell, Lorie Garcia, Dixie Garr, James Gibbons, Dr. Sam Haddad, Stanley Hiller, Jr., Yvonne Jacobson Olson, Earle and Nicholas Jones, Guy Kawasaki, Grace & Gay Kennerson, Eugene Kleiner, Bill Lohse, Mike Malone, Bob Marshall, Stanley Moniz, Jerry Morissette, Craig Newmark, Kim Polese, Carol Sands, Aram Saroyan, Grace Slick (you rock!), Judy Stabile, Jill Tarter, John Warnock, Suzan Woods, Margaret Wozniak, Ian Patrick Sobieski, Ph.D. and all three generations of the Baird family — you are all responsible for giving a flavor to the past and the future that I hope shall remain with the readers of this book for a very long time.

Brenna Bolger, PR woman extraordinaire (PRX Inc.com), deserves a special kudos for helping with lining up some very special interviews, thank you, Brenna! A warm debt of gratitude to the late Lawrence Hollings who taught me the importance of dreaming the future. Thanks to Charles Kittleson of Vacuum Tube Valley (vacuum-tube.com) who helped fill me in on much of the Valley's early technologies and loaned me countless vacuum tubes for my photo shoots.

And an especially heartfelt thanks to Scott "Scatman" Clark, my friend of the ages, for the endless use of his Mac Central Mediabase. Scott's tremendous patience as a wafer wrangler on the beach shoots was endless and he was especially understanding when the ocean came in and took his wafers back to the sea from whence they came.

Speaking of sand and silicon wafers, thank you Satya Sreenivas for the first half of the title. Satya said, "Call the book Sand Dreams." And so it goes.

I conducted hundreds of interviews for this work, but relatively few made it in, so a very big thank you to the folks who gave me your time. Although space restraints wouldn't allow for the use of all of your interviews, believe me when I say I heard your combined voices when writing the texture for *Sand Dreams & Silicon Orchards*. There are so many people to whom I would like to send my heartfelt thanks, many of whose names I don't even know. In writing this book, I wandered aimlessly for a few months, picking up pieces here and there and ending up in places like the soon-to-be-demolished Walker's Wagon Wheel and Ricky's where I just sat and talked with people, soaking up atmosphere. And there were all the rangers and keepers of open space and private property who let me range their properties to shoot thousands of photos.

Speaking of photos — Debora Cartwright and the entire gang at the Dark Room in San Carlos were outstanding at processing the rolls and rolls of film I just kept on shooting. Their ongoing advice and constant source of support is truly a gift. A big kudos to my accountant Mike Vivrette who found all of the deductions in the research for this book, you are a demigod.

A special thanks to Pat Hagen, my personal ethicist, who steered me away from the politics involved in writing this book. And a warm remembrance of Warren "Dad" Mack who instilled in me the nuances of objectivity, the beauty of brevity and the complexity of commentary. A lifetime of thanks to Biko Richards who taught me that life is not worth living if you don't fill each and every day with the curiosity to solving its mysteries.

And a huge debt of gratitude to my niece Simonne, a wonderful source of inspiration, who spent her early childhood in Silicon Valley and who may one day return home to become a dot-com millionaire.

Photo by Sally Richards

Photo by Sally Richards

INTRODUCTION

When I was accepted into graduate school at Stanford, I didn't have to think twice. Michigan, Georgia Tech, Purdue — they were all immediately forgotten. I'd never even been to California and why I was so certain that this was the place for me to go is a mystery still. Perhaps it is because the myth of California is the myth of America — a divinely beautiful land, full of strength, vitality and possibility, with natural splendors like Yosemite and a people who lead the nation that leads the world. Even in the mountains of Virginia, where I lived, the myth of California pervades.

Arriving in the Valley in my tiny Honda CRX, I marveled at the little differences that I now take for granted. The cars are not Rolls Royces but they are uniformly new and nice — there is a culture of high quality in everything from the manicured lawns to the artsy cinemas. At the time, I couldn't spell entrepreneur, much less relate to what one really was. But slowly the culture would break into even my thick head.

I lived on Addisson Street and about four blocks away was the garage where HP started, a historic plaque outside announces its significance. I shared a beer with Jerry Yang and his roommates in his Forest Avenue apartment and heard about the computer stuff that was so distracting him from his thesis a year before he started Yahoo! I met a woman who, like me, was in her mid-20s, though she was a millionaire incongruously early and plotting a year of round-the-world travels.

This place was different! From the cars to the movies to the people, it slowly changed who I was and what I thought was possible. And now I am part of that story, a small part to be sure. It still amazes me how people arrive in Silicon Valley and make a piece of it their own by building a company or developing a technology.

As a venture capitalist, my group makes risky but potentially rewarding, early investments in companies, a tradition that has long gone on here in the Valley. I meet a lot of entrepreneurs, and without exception they are united by a culture in the Valley that pervades a sense of optimism and possibilities. People arrive from all over the world to be part of this culture; *Sand Dreams & Silicon Orchards* tells the story of how all this came to be.

Ian Patrick Sobieski, Ph.D.

Managing Director, Band of Angels Fund, LP

April 28, 2000

Photo by Sally Richards

 dawn

"I am very sad; my people were once around me **like the sands** of the shore — many, many. They have gone to the mountains [died] — I do not complain; the antelope falls with the ARROW."

— An aged American Indian near Mission Dolores

CHAP

Photo by Sally Richards

CHAPTER ONE

A new dawn was just beginning in California as a people who had lived by the land's rules for nearly 11,000 years were about to meet the destiny that would soon land upon their shores. In myth, they knew others existed as they had spent lifetimes looking out over massive oceans for signs of the "others." Surely, one day, they would come. They must have had different reasons for hoping or fearing others were living just on the other side of the horizon. They may have wondered how others dressed, what they ate, how they lived, if they spoke the same language or had the same dreams. They would soon solve the mysteries created by their own imaginations and fed by human nature's need to survive and conquer. This was the dawn of a new civilization, one that would eventually become Silicon Valley.

It's difficult to imagine this now bustling valley was once a serene place where nature and her seasons were the only elements dominating day-to-day life. At one time, Native Americans lived side

This Ohlone hunter was sketched by an artist with the La Pérouse Expedition in Monterey Bay in 1786. Skillful hunters, the Ohlone stalked larger animals with obsidian-tipped arrows and sinew-backed bows.
J. Culleton, Indians and Pioneers of Old Monterey, 1950

by side with nature, and animals such as elk, pronghorn antelope and bear roamed in great numbers on the valley's floor. The days were crisp and clear, and from surrounding mountain peaks one could clearly see all the way to the lapping waves of the Pacific.

This story begins in a very controversial time in the world's history. Each people involved — the Spaniards, the Mexicans, the Americans, the Native Americans and others — had their own standards and laws, written and unwritten. Unfortunately, few saw eye to eye. It was a wild and tragic era when people did everything humanly possible — all in the name of love of country — to get ahead, to plant the flag, to gather wealth. Others only watched in disbelief as their lives changed and everything as they once knew it came to a halt.

California, one of the last desirable places on this continent untouched by the monolithic wave of European conquest, was a land teeming with groups of indigenous peoples until the 1500s. In 1492 there were believed to be 10 million Native American Indians in the United States. In California, many tribes had fluid social relationships, but Native American nations did not exist there as they did in other parts of the country. Current history dictates that the Costanoans, or the Ohlones, the Bay Area's first Native Americans, were also the first people to inhabit the Bay Area. For the most part, they worshiped the earth and her creatures and led fairly peaceful and fulfilling lives.

The Ohlones didn't have an easy life, but it was the life they had chosen, one where they were free to worship, love and live however their wants and needs led them. The fertile ground had yet to be tilled, and the rivers that ran from the mountains into the bay and ocean provided them with drinking water, oysters, waterfowl and fish. With rare deviance, Ohlones were born, married and died within 50 miles, or much less, of their birthplaces. The people shared their land with the eagles and grizzlies and trusted that the earth would provide for millennia. And why not? Their people had gone on before them for more than 11,000 years.

Because the Ohlones supposedly had no governing written language (the Native Americans in this region had more than 50 different dialects), no recorded history has been found or understood that will completely tell their story.

Throughout the 16th and 17th centuries, representatives of the Spanish government were systematically destroying other cultures while conquering the indigenous peoples and their land, a continent they referred to as New Spain, now known as Mexico. They left many of the empires completely devastated in their wake. With little or no immunities to European diseases, many thousands of natives died. Thousands more perished under the pure brute force brought against them in the name of God and country. Genocide had taken place in New Spain, and by the mid-18th century, the next wave of tragedy was about to land on California's rugged coast, a landscape that, by its wild nature, had always protected her from change.

Imagine a world where news traveled so slowly that by the time it crossed an ocean it had been revised and embellished at least a thousand times. With life spans relatively short and blights as simple as scurvy killing entire shiploads of explorers, dreams of adventure and wealth were the only elements overriding the fear of the unknown for individuals and empires pushing beyond their boundaries.

As always, mythology and legends were powerful aphrodisiacs for sailors, mercenaries, dreamers and the armies wildly roaming the world and keeping their ears to the ground for rumors of gold-paved streets and mounds of emeralds. As is the custom for today's venture capitalists, queens, kings and the wealthy in general kept funding these journeys as long as there was opportunity for a payoff. One could say that the investors of the New World explorers were the prelude, the great ancestors, to today's Silicon Valley venture capitalists. Thus, these rogues of the ocean were the role models for today's entrepreneurs. Unfortunately, unlike it is today, relatively few found fame, wealth or ease of life.

It was about 1510 when these explorers began to hear tell of a story that had spread like wildfire. The name of the fantastical story was *Las Sergas de Esplandian,* written by Garci Ordonez de Montalvo. The story told of a mystic place referred to as California, a magical land populated by Amazon women who used males for breeding purposes alone, feeding the excess to trained griffins. Coincidentally, the only metal known in this sunshine-filled world was gold.

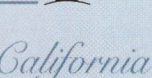

California

I wish to tell you about the strangest thing ever found anywhere in written texts or in human memory. I tell you that on the right-hand side of the Indies there is an island called California, which was very close to the region of the heavenly Paradise. This island is inhabited by black women, and there were no males among them, for their lifestyle was like that of the Amazons. These women had powerful bodies and courageous hearts, and they were very strong. The island was made up of the wildest cliffs and the sharpest crags found anywhere in the world.

On this island called California, with a great multitude of wild animals, are many griffins that are not found in any other place. When the griffins give birth, the women cover themselves with thick hides to snare them. They bring them back and raise them in their caves.

Their armor is made entirely out of gold — which is the only metal found on the island — as are the trappings on the fierce beasts who they ride once they are tamed. They have many ships that they use on their raids to carry away the men they seize and feed to their griffins. On occasion, they keep the peace with their male opponents, and the females and the males mate with each other, from which many of the women become pregnant. If they birth a female, they keep her. If they birth a male, he is immediately killed. They are set on reducing the number of males to a small group that they can easily rule over.

— *Las Sergas de Esplandian*
Garci Rodríguez Ordonez de Montalvo
1510, Spain

Photo by Sally Richards

Rumors of wealth beyond New Spain plagued the Spaniards who had effectively bludgeoned Mexico into submission. The outpost of San Miguel de Culiacan marked the most northern Spanish conquest, and no one knew what lay beyond the dangerous terrain of the Sierra Madre. It was 1536 when Captain Melchior Diaz sent out his men to find what they would. What they fetched were three Spaniards and a black slave, all nearly dead but hallucinating wildly of gold and riches.

They told a winding tale of their incredible trek beginning in 1528 when a shipwreck of nearly 1,000 people washed ashore in Florida. Not many survived. Most were captured, enslaved and killed. These four men had seen many cultures and tribes on their long walk through Florida, Texas, Arizona and deep into Mexico where they were finally found. Although they were obviously not any richer for their journey, they told accounts of wealth, mounds of emeralds and mountains of gold.

All of these stories added to the frenzy of Diaz's imagination that constantly churned with thoughts of the wealth waiting for him to claim. Things moved slowly in New Spain, but in 1540 the conquistador was put in charge of an army that would meet up with the ships of Indian slaves, Spaniards, cattle and friars. His huge party was soon on its journey to find California. Although Diaz was the first Spaniard to see California, he had no idea what he was looking at. On his way to California, he was waylaid and ended up on the Colorado River where he was able to see the Promised Land. He soon thereafter disappeared from California history.

In 1542 a ship under the control of Spanish Captain Juan Rodriguez Cabrillo, a Portuguese explorer funded by Spain, supposedly made the first discovery of Alta (upper) California, an exploration that included Monterey. Cabrillo named what is now Monterey Bay *La Bahiá de los Pinos,* The Bay of the Pines. The explorers mapped as far north as the Oregon coastline before returning home with news of their discoveries. The explorers returned home without mounds of jewels or even a single trained griffin for all of their efforts, but they did make it back (sans Cabrillo who had died on the trip).

Time passed and political climates changed in Europe. Other countries began to pursue exploratory interests and soon mercenaries and military captains were receiving the go-ahead to forge forward with the flags of their mother countries.

It was 1578 when pirate Sir Francis Drake, who wasn't actually knighted until after his return home from this journey, entered the scene. Drake, who received his funding from Queen Elizabeth I, headed off to find a passage that would give the British a trade foothold in Asia. His dreams of riches were realized as he found glittering bounty in nearly every ship he forced himself upon. His final acquisition on the Bay Area coast was the treasures and crew of the Spanish ship *Cacafuegos.* It was this last bounty of confiscated cargo that caused his ship's hull to bulge and groan. After he loaded the more than 1,000 bars of silver and hundreds of pounds of other metallic bobbles, the extra weight of his booty nearly destroyed his ship. Drake sidled his greedy ship into a cove off of the California coast for repairs.

Drake, a sometimes charming pirate, freed the crew of the *Cacafuegos* to return to their ship and even gave them a letter in case they were delayed any further by other ships in his fleet. In the letter, he gave his word that if anything else was taken from the ship by his fleet, he should ask his men to pay double its worth, which he promised to pay in full upon their next meeting.

Where Drake actually landed is still somewhat of a controversy. Some say he landed in what is now Drakes Bay, north of San Francisco. Others say that he landed near the bay where San Quentin Federal Prison is located, in Marin County. A journal written by an eyewitness claims that the locals showed themselves and brought gifts. Drake, in turn, gave them clothes to cover their nakedness. Drake asked the local Miwok to give up their land in the name of the Queen; they hadn't ever been asked to do such a thing. They appeared to genuinely enjoy Drake's company, humored him and then they fed the hungry army of men.

The Miwok people were eager to please their new guests, who were probably the first white men they had ever seen. They even gave the seemingly generous Drake a crown "made of knit-work wrought with feathers of divers colors, the chains being made bony substances," which he immediately accepted in the name of England and her queen.

Drake stayed for five weeks and called his new land Nova Albion. He placed a brass plate there calling it such and claiming it in the name of England. But a plaque isn't

worth much without the paper to prove it. Since there were no actual contracts between nations, or even an understanding of what a land transaction implied, the country, and apparently her people, were still considered unclaimed.

In 1602 Sabastián Vizcaíno, an explorer sent by Spain, rediscovered the Monterey Bay. He made note of snow capping the Santa Lucia Mountain Range, as well as the chore of having to break ice to obtain fresh water. As every explorer is apt to do, he renamed his rediscovery. *La Bahia de los Pinos* was now dubbed Monterey, in honor of Viceroy Conde de Monterey who had sanctioned his voyage. The party spent 18 days in the harbor and were the first, on record, to encounter the bay's Ohlones.

The next explorer to rediscover the Bay Area was General José de Galvez of His Majesty's Supreme Council of the Indies. This ambitious sailor fervently hoped to develop all of New Spain and California. At the time, Spain was reeling from the Seven Years' War and resources were placed only where absolutely necessary. So, to better advocate his position, Galvez spread panicked rumors, some built on partial truths, of Russia taking all of California in her own name.

The Seven Years' War had ended with the 1763 Treaty of Paris. This document would temporally seal England's rights

The English navigator and privateer Sir Francis Drake "discovered" California in 1579. Having plundered the settlements and ships along the west coast of the Americas, Drake was blown off course as he attempted to cross the Pacific. Seeing white cliffs along the coast of Northern California, he called this unknown land New Albion and claimed it for Queen Elizabeth of England.
Frank Soulé et. al., Annals of San Francisco, 1855

The Franciscan missionaries often traveled to the native villages to baptize the sick and the children. In time they persuaded the villagers to take up residence at the missions where they received food, clothing, shelter, occupational training and religious instruction.
Frank Soulé et al., Annals of San Francisco, 1855

to North America. About this time, Charles III banished the heavy-handed order of Jesuits of New Spain, replacing them with the more favored members of the Franciscan order.

Galvez's greedy dreams were realized when he received orders from Spain to found a military presidio in Monterey. First, he ordered the captains of the ships *San Carlos, San José* and *San Antonio* to organize a presidio in San Diego. Galvez was heady with power, as he would be granted absolute rule over the expedition and then would serve as governor of all of Alta California.

Nearly 200 years had passed since Vizcaíno's visit to Monterey, but in 1769 when Galvez's ships sailed, the land of myth and fortune never would be the same again. The journey began as the *San Carlos* overshot San Diego, ended up near Santa Barbara and had to backtrack to meet up with the *San Antonio*. When she arrived in San Diego, her sailors were plagued with scurvy and the group ended up building a makeshift hospital on the beach. Many of the crew on the *San Carlos,* who were helping to nurse the others back to health, were also brought down by disease. The *San José* was of absolutely no help. It was heard from briefly, tried once again to sail for San Diego, but eventually was sunk, pirated or mutinied before disappearing altogether.

The *San Carlos* and *San Antonio* ended up losing half of their men. Meanwhile, the part of the expedition led by Gaspar de Portolá that was traveling by foot came along and continued pushing north. The group included Father Junípero Serra, the president of the Franciscan Baja missions. Serra's job was to establish the newly founded country with missions and colonize the native population. Spain's cold and systematic way of settling a country partially depended on the priests' success in taming local natives before assimilating them. For many innocents who stood in Spain's path, resistance was futile. As the volunteer converts became slaves of the church, they either became passive and productive, or died rebelling.

This colonization process included three elements that were intertwined, each weighing heavily upon the other for total success. The first was the presidio, a military garrison providing the heavy hand of the law. The second was the pueblo, a community of settlers who would form an agricultural base and provide the presidio with food and all of its necessities. The missions, which Serra was now in charge of bringing about, would bring the natives together, hold their property until they were proclaimed citizens and basically enslave and exploit them for free labor until Spain's work was done.

"So you think we can make the venture a success?" asked Galvez of Serra one day before embarking on the journey.

"Surely," said Serra. "It is God's work to carry the cross of the holy faith into the wilderness, and He will go with us; can you not hear the heathen calling us to bring them the blessed Gospel? I can see that I have lived all my life for this glorious day."

On their journey, fathers Serra and Juan Crespi made note of village after village of Native Americans along the rugged coastal range.

"The Indians were amazed," said Mark Hylkema, archaeologist and Native American authority. "And they fed them, which is a tradition that I continue to experience to this day. In Santa Barbara they encountered some large communities, every half-mile there were villages of about 1,000 maritime folks who fished from canoes on the sea. They continued on, but kept encountering this incredible hospitality and were given fish and food. The Indians danced into the night and they couldn't get any sleep. So they finally said, 'thank you, but no thank you' and continued on to the Santa Lucia Mountains."

The explorers continued past Santa Barbara, purposefully avoiding American Indian villages. There were resources for the travelers and they hunted bear in the area now known as Los Osos (the bears).

"They continued north until they got to Point Lobos, and they looked at their maps based on Vizcaíno's observations, looked at the area and said, 'this isn't it,'" said Hylkema. "Meanwhile, they were supposed to meet a ship, but the ship hadn't come, so they made a cross that could be seen from the sea and they left a note in case someone should come. They also ran out of supplies and took a vote whether they should go on — you never take a vote on a military expedition — this was serious."

In the the fall of 1769, the group voted to continue north. Luckily, they found a huge village located along what is today the Pajaro River in Watsonville. The villagers were frightened when they first encountered the ghost-like men. The women and children bolted for the hills while the men stayed behind. "The men ran forward and stuck their arrows into the ground," said Hylkema. "It was an act of defiance, a marker of traditional warfare. Sergeant Ortega, a leader who will live on in California's history, got off his horse, walked up to the arrows and pulled one of them out of the ground. The Indians began clapping; he did the right thing. They avoided conflict and the Indians brought them into the village. The women and children came back from a celebration and promptly fed their guests."

The explorers continued north, looking for charted landmarks that would help them on their journey to find Monterey.

"At Santa Cruz, all of the grasslands were burned and there were no Indians around," said Hylkema of what the explorers witnessed on this leg of their journey. "And they were concerned because they didn't have any forage for their animals. They wanted to find an Indian guide, but could find no one. What they didn't know was that the Indians burned the lands every fall to increase the grass seed harvest. They also burned the oak woodlands annually to eliminate the lower sapper shoots so trees would put more energy into producing acorns. They burned land close to the villages and got improved vegetation in the spring that attracted more grazing animals such as elk and deer. The women of the villages pruned the bushes to harvest straighter basketry. They completely managed the landscape; these actions were the early roots of agriculture and livestock management that every culture on earth goes through."

California Native American baskets
Photo by Deborah Lohrke
History San José

The Valley's Native Americans relied on acorns from the area's oak trees as a primary food source.
Photo by Sally Richards

Evidently, the Pajaro Indians had given them berries, but it had taken a few days for their bodies to absorb the healing vitamins. In honor of their speedy recovery, which they entirely attributed to prayer, they named the area *La Salud* for good health.

"They finally got to Montera and a place now called Martini Creek, a state park near Devil's Slide," said Hylkema, "and they found a cove of mussels, which they named *Punta de Las Almejas,* point of mussels. They hunted wild geese that used to inhabit the shore before the wetlands were drained, and climbed to the top of a ridge to shoot deer. What did they see on the other side? Stretched out before them was the San Francisco Bay. They also saw the Farallones and Point Reyes to the north, both recognizable nautical landmarks for them."

The explorers knew they'd overshot Monterey and there was no ship in sight. The group was nearly out of supplies, so after a cursory exploration of San Francisco, the weary men headed south. They went along the interior bay shore through Portola Valley and camped by Fransanquito Creek in Palo Alto, near a tall redwood in the area where Stanford University now stands.

"There were Indians everywhere," Hylkema said, describing what the explorers found only 230 years ago.

By this time, without the generosity of the natives who had been consistently feeding them throughout their journey, many in the group fell ill. With the weather dismal and the men weak of spirit and body, the group set up camp at Wadell Creek near Año Nuevo. The men were dying of scurvy and receiving last rights from the priests. Then, it surprisingly began to rain and it seemed as though the journey would come to a close sooner than anyone imagined. Two days later everyone was fine again.

"The Indians spoke Spanish words perfectly upon hearing them once, and one of the priests commented about how charmed he was by these malleable Indians. One of the officers noticed how the Indians exhibited typical human behavior in the way that they maneuvered to have the advantage, by keeping the soldiers to their left at all times, which was their bow side. They camped in Palo Alto and Ortega took six soldiers to reconnoiter Contra Costa, which means 'the other side.' They went to San Jose, Oakland, Richmond and

Hercules, and what did they see? The Carquinez Straights, a vast tule wetlands swamp. They couldn't get across, so they went back to the creek where Father Francisco Palou was writing in his journal and describing how there were bears everywhere, and how huge they were."

After a bit more exploring, the group decided to return to San Diego to regroup. At the time, the Bay Area was wild and quite stunning. San Jose and its surrounding land was a cornucopia of ecological zones. Tidal bay marshes surrounded the bay (but later were filled with silt from the hydraulic mining of the Gold Rush era.) It was a vast tule reed marsh wetlands where Native Americans fished from their reed boats in marshes thick with wildlife.

The Ohlones, using the flesh of animals for food, the bones for tools and jewelry, and skins and feathers for clothing and other utensils, collected ducks in their nets and harpooned sturgeon, salmon, stingray and harbor seals from the waters that flowed freely throughout the Bay Area. Elk drank from the shore and sea otters played in the bay's waters until the 1830s, when they had already started to become a rare sight.

Along the edge of the marsh was a huge willow thicket that reached all the way from Alviso to Redwood City, and there the grassland prairie met the tidal marsh habitat. The band of thicket stretched almost to San Jose where the topography began rising and gave way to the oak woodlands that formed the belt around the grasslands. From the woodlands rose the base of the foothills with vernal pools where water percolated up through the ground. A thicket filled with juniper and manzanita grew all the way from New Almaden into Palo Alto. In those days the oaks were enormous. Explorer George Vancouver noted one oak tree in the Sunnyvale area that measured 15 feet in girth and claimed there were many that were larger. In odd contrast, many of the largest oak trees remaining in the Valley are safely growing between Page Mill and Sand Hill roads in Palo Alto, an area where much of the world's venture capital is acquired... cash sometimes referred to as "seed" money.

Tribes claimed about 10 square miles each and had access to resources as the seasons changed. There were marriages and friendships between the tribes, so what they didn't have, they acquired by trade. By following seasonal cycles of abundance and scarcity, and developing a market economy using shaped shell beads, they wanted for very little.

Upon Spain's return in 1770, Captain Gaspar de Portolá established the Monterey Presidio and soon disappeared from California's history. His replacement was Pedro Fages.

Meanwhile, Felipe DeNeve, a Spanish officer, was under a royal order to set up Pueblo San Jose along the Guadalupe River. According to rules governing Spanish providence, the community was illegally close to Mission Santa Clara. The two had different goals, each in direct conflict with the other. The mission's goal was to bring the Native Americans in by voluntary enticement, and there was little incentive for the natives to give up all they had to be sometimes quite brutally beaten and have God, someone else's God, foisted upon them.

"The Indians were attracted by one thing," said Hylkema, "access to new material goods. They also built the missions right in the middle of their gathering grounds, so what were they going to do? They couldn't go en masse to their neighbors because territory was a well-established matter among the tribes, and those situated where the missions were had no matter of recourse other than to join the church. They did welcome the Spanish because of the new products and commodities; unfortunately things soured pretty quickly at Mission Santa Clara because the Indians killed a mule, and that was forbidden. They were hungry, so they killed it to eat it. There was an immediate detachment of soldiers sent out from the Presidio Monterey who fell upon the Indians in the middle of the night on the Peninsula. They were roasting the mule meat. They fought back and the soldiers killed five of them. They took the leader back to the mission where they whipped him. Of course the priests tried to plea for clemency, but DeNeve was a bit more hardheaded about it."

Father Serra was incensed by the entire situation and sent many letters of complaint to the viceroy about DeNeve. The Native Americans didn't want to go to the mission when they could instead go to the Pueblo San Jose. At least at the pueblo they could be paid in cloth and metal and odds and ends that were unique. Why should they go through the trouble of converting? Mission Santa Clara was at ends with the Pueblo San Jose from the very beginning.

This photograph of Mission Santa Clara depicts the mission as it was in the 1850s. The fifth of the Santa Clara Mission churches, this building was dedicated in 1825. After the mission was secularized in 1836, it served as the parish church. In 1851 the property was transferred to the Jesuits who founded Santa Clara College. The college building can be seen on the right. The adobe bell tower fell in the earthquake of 1838 and was replaced by a wooden tower.
Sourisseau Academy, SJSU

The Franciscan priests' ultimate job was to turn the Indians into *gente de razón,* men of reason. Within 10 years of their "reasoning" training, which included heavy labor and little food and resources, the priests would then make them citizens of Spain and bestow land to them. Many years later at Mission Santa Clara, of all the Native Americans that went through the process, only three received land. One of the three was Lope Inigo, who at one time had owned what is now Moffett Field and beyond. He died in 1864 and was buried under what is now the intersection of highways 237 and 101.

Needless to say, the California missions, 21 in all, were not successful in colonizing the state. The mission Native Americans were separated from their families and often were brutally treated for the slightest offense. Mortality was high, especially among children and women. Unlike the Ohlone's natural way of ingesting herbs for prenatal care, the missions practiced prayer instead and great numbers of babies were delivered stillborn. Compounding the mortality rate were also the documented contributing factors of rape, venereal diseases and starvation.

Abortion was practiced by female Native Americans who were forced to continue manual labor well into their pregnancies. There were also cases of mothers smothering their babies and small children, probably done with the forethought of sparing them from such a dismal life. A war of psychological torment and physical fear had been waged against the Ohlone people and they were losing what little leverage they had. Lack of nutritional food and even the clothes provided by the missionaries contributed to intensifying the diseases they were already susceptible to. Their spirits were low as many died around them. A people who they had welcomed now totally enslaved them for the purpose of manual labor, all in the name of colonial ambition.

Twice annually the Native Americans at the mission were allowed to go home. This was a last-ditch effort to keep the Ohlones controlled as well as possibly encourage them to bring other family members back to the mission to convert. It also allowed the Native Americans, who were unable to acquire enough food to sustain themselves in a healthy way, to forage for food away from the missions.

One of these "vacations" was observed in 1816 by visitor Otto von Kotzebue who wrote the following: "I, myself, have seen them go home in crowds, with loud rejoicings. The sick, who cannot undertake the journey, at least accompany their happy countrymen to the shore where they embark and sit there for days together, mournfully gazing on the distant summits of the mountains which surround their homes. They often sit in this situation for many days, without taking any food. So much does the sight of their lost home affect these new Christians. Every time some of those who have permission run away, and they would probably all do it, were they not deterred by their fears of the soldiers."

The land's characteristics were changing as more settlers began pouring into the area. Elk, bear and deer were quickly killed for sport and replaced by domesticated cattle and horses. Vast numbers of domestic livestock damaged the native landscape. The Native Americans, who had always relied on the land, were having a difficult time finding food. Those facing hardships stole cattle, but if not completely stealth-like in their "crimes," would be hunted down like animals and executed.

Following Mexico's declaration of its independence from Spain, it took Mexico 10 years to win California over. Spain had stopped funding its armies and the missions, and soon Californians decided they could be no worse off with Mexico than they were with Spain, probably better. Spain was ousted and a new transition took place. California opened the trade channels that Spain had made illegal and the state began prospering.

In 1834 secularization was foisted upon the missions by the Mexican government and the land was redistributed to ranchers in the form of large Mexican land grants. But nothing lasts forever. Californios, Californian-born Mexican settlers, began looking at life as the rebels would during the American Revolution.

"California was on a path and it wasn't going to stop," said Lorie Garcia, historical consultant and author. "This was a group of bright young men, native-born Californian citizens of Mexico. They were asking, 'Why should we be governed by a place 1,000 miles away? We should govern ourselves.' It was a complex couple of decades. The tallow and hide trade was really picking up, and instead of money — Mexican money, American money, European money, it was all being used there — Californians were trading hides that become known as California bank notes. The hides were shipped back East, processed and sold back to Californians after they were

Early Native Americans at Mission Santa Clara
California History Center

The Valley's early wheat farmers
Redwood City Public Library Local History Room

The influx of miners and traders into the region severely depleted the population of deer, elk and bear in the Bay Area.
Redwood City Public Library Local History Room

made into leather goods. They could get $2 in trade for a skin, but would pay $8 for a pair of shoes, the shoes their hides were made from."

The year 1846 marked the United States' conquest of California. Shortly thereafter, in 1848, gold was discovered in California's interior area of Sutter's Fort, and destiny moved on full-speed ahead.

There was also no turning back for the Native Americans who called the state home. As more immigrants moved into California at an alarming rate, native people were forced to move farther into the mountain ranges. Food became scarce as deer, once traveling in herds of 1,000, were hunted down by the influx of miners and traders. Many Native Americans fought to reclaim what little land was left to them, often having to take domestic animals to keep their tribes from starving.

Because many of the county's ranchers were calling the Native Americans cattle thieves, the state government declared them the enemy and offered $5 per head and 50 cents per scalp bounty. In 1854 in California alone, the federal government reimbursed the state of California $1 million that was paid out to Indian bounty hunters. Many coming into California who could not make a living mining gold became involved in the wholesale slaughter of the indigenous men, women, children and infants of California. Often the human predators would choose to keep women and children alive and sell them to local ranchers.

In 1850 the U.S. government assigned three Indian Commissioners to establish treaties and reservations with California Indians. On September 16 Commissioner Adam Johnston noted the following as his findings:

"Almost the entire tribe of coast Indians have passed away. Of the numerous tribes, which but a few years ago inhabited the country bordering on the bay of San Francisco, scarcely an individual is left. The pale-faces have taken possession of their country and trample upon the graves of their forefathers."

In an interview with Johnston, a very aged Native American near the Mission of Dolores [San Francisco] said, "I am very sad; my people were once around me like the sands of the shore — many, many. They have gone to the mountains [died] — I do not complain; the antelope falls with the arrow. I had a son. I loved him. When the white man came he went away; I know not where he is. I am a Christian Indian; I am all that is left of my people. I am alone."

Throughout the years of discovery of this land and its changing political climates, the California Indians were wooed, controlled and finally nearly wiped out. Today, there are now believed to be only 150,000 native peoples in the state of California. For each new civilization's dawn, another era must end. And thus begins California's next movement in time.

Photo by Sally Richards

Photo by Sally Richards

Photo by Sally Richards

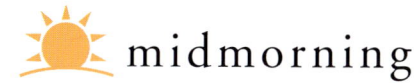 midmorning

"I always have to dream up there **against the stars.** If I don't dream I will make it, I won't even **get close.**"

— *Henry J. Kaiser*

CHAPTER TWO

In 1848, as gold fever intensified, so did the entrepreneurs who were making money by selling the byproducts of the Gold Rush. While many came all the way west to find fortune, relatively few actually did. Some miners were pulling $100 to $700 of gold every day from the creeks of interior California, but as the easy gold was plucked from the silt, the miners needed to burrow deeper into the mountains to find what they were looking for. As they did so, more miners and mining companies filled the area to stake claims.

The world's vagabonds and hard-working men landed side by side at San Francisco's harbor. The once relatively small city was growing by tremendous strides and taking the entire Bay Area with it. Immigrants often arrived broke with the goal of raising enough coinage for a gold pan, a shovel and passage to 49er country.

An early Bay Area taxi
Redwood City Public
Library Local History Room

Redwood City
dock workers
Redwood City Public
Library Local History Room

Turn-of-the-century
Bay Area
Redwood City Public
Library Local History Room

Action in downtown
Redwood City
Redwood City Public
Library Local History Room

Newly rich miners, with the bravado of a thousand drunken kings, wandered the city looking for creative ways to blow their fortunes on whiskey, women and another chance at getting richer. Also lurking about were grifters and opportunists conniving ways to part the miners from their cash. The harbor town was growing, and it stretched out its wood and brick by blocks every day. As the city sprawled, the need to feed it with lumber increased.

> "I've labored long and hard for bread, for honor and for riches, but on my corns too long you've tread, you fair-haired sons of bitches."
>
> — *Black Bart*, notorious San Francisco bandit of the 1870s, left this note in a San Francisco strongbox after having robbed it. With the increase of disappointment and desperation came the crime that would follow.

The extreme lengths that entrepreneurs would take to make and keep their money were becoming legendary. Characters poured into all corners of the Bay Area to make their mark upon history. All came because this place of mythic proportions had whetted their greedy appetites with fantasies no less in grandeur than Diaz's thoughts of gold-paved streets and mountains of emeralds. And, as throughout California's history, everyone expected the land to produce their dreams and pleasure their souls.

One of the major industries affected by San Francisco's growth was lumber. Long before Silicon Valley's elite began building mega-estates in the mountains of Woodside, it was rough terrain that taunted only the most ingenious problem solvers. How would they answer the challenge of cutting into the thick forests, severing trees, processing wood, getting the lumber down the steep, untamed ridges and finally off to market?

A Growing Business

The one man who did a great deal to define the lumber techniques that would shape the future community of Woodside was Dr. Robert Orville Tripp, a jack of many trades and ambitions. Tripp, originally a dentist from New England, had moved to California to find economic opportunities. His story was typical of the throngs of settlers who had left their businesses on the East Coast and brought their marketable skills to California.

Tripp had formed a partnership with Mathias Parkhurst and it wasn't long before they were making a profitable living bringing their lumber to market faster and more efficiently by working smarter, not harder. It's the same story today, being first to market is still half the battle.

According to Gilbert Richards' book *Crossroads,* Tripp is credited with developing a method that supplied San Francisco with wood for its wharves (much of it is still there). Tripp cut logs from Woodside's vast forests and dragged them down Redwood Ravine (a route credited to Tripp's discovery) by oxen to the water of the embarcadero in Redwood City. There, the logs would be lashed into huge rafts anchored to shore until the tides shifted, and then the journey would continue on either to San Francisco or farther south to Alviso to build San Jose.

The Redwood City embarcadero was a wild place during those times. It didn't have as many immigrants pouring in as San Francisco, but it gave the City by the Bay a run for its money. The streets were lined with more than a dozen breweries and it wasn't a place for the weak of heart.

It was a typical week in Redwood City when 24 ships would arrive to the dock to take away 50,000 board feet of lumber, more than 1 million shingles, nearly 200,000 fence posts and upwards of 200 cords of firewood. Soon boats were being built on the shoreline property and fruits and vegetables grown in the South Bay were being shipped to San Francisco and beyond. Blacksmith and wagon repair shops began to pop up along the lumber roads from the hills to the wharf.

Tripp was making quite a name for himself and for the budding town of Woodside. Knowing he had to provide niceties and necessities to his workers at the mill, he and his partner teamed up to open a general store (the Woodside Store still stands). Although he had a mercantile license, he did not have the county's permission to sell

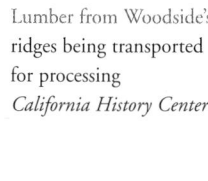

The Woodside Store
Redwood City Public Library Local History Room

Lumber from Woodside's ridges being transported for processing
California History Center

Wood from the coast ridges was hauled down and pushed off a chute onto a ship below for transport.
Redwood City Public Library Local History Room

whiskey by the shot. He soon found that as a dentist, his power to write prescriptions was also his license to sell whiskey, and he began "prescribing" shots of whiskey to soothe his patients' ills. There's no doubt that Tripp was a popular resident — so popular that he served on the first Board of Supervisors in San Francisco in 1850, and as the public administrator in San Mateo County in 1850 and 1863. By then, Tripp was Woodside's banker, grocer, dentist, merchant, bartender and postmaster.

At that time, what is now known as San Mateo County was part of San Francisco County. In 1853, San Mateo County broke away and was coming of age. Scores of sawmills dotted the mountain ranges from Woodside to Davenport. Those who had money purchased mills and redwood groves in hopes of cashing in. Many of the coastal lumber mills loaded logs directly into a chute and onto ships docked in Pescadero.

When the Almaden Quicksilver Mine was a booming city
Redwood City Public Library Local History Room

By the latter half of 1858, the harbor town of Redwood City was as raucous as one would find anywhere. Livestock was left to graze in the town, and bars and hotels did quite a brisk business. There were even shootings in the streets and crime, although nowhere near what it is today, was on the rise.

QUICKSILVER BUILDS AN EMPIRE

Meanwhile, there was another fortune to be made in the South Bay from another natural resource. Twelve miles from Pueblo San Jose, mercury (then known as quicksilver), which had been in high demand in other parts of the world, was now so in the Gold Rush Country. The quicksilver mine that would be the mother of them all was located right in San Jose's New Almaden. Not everyone saw the value in the metal that had no stability and was far too soft to make coinage, jewelry or other bobbles of worth. According to Jimmie Schneider, New Almaden Mine tributor and author of *Quicksilver:* "Only the alchemists valued it, and they were not considered to be practical men. They believed, by rumor and fancy and musty old books, that quicksilver was a chief component of the Philosopher's Stone, which reportedly could turn waste materials into gold."

And that it did, to a degree. The New Almaden Mine was a wonderfully rich cinnabar mine. Mercury and sulfur combine to make cinnabar, a mineral that varies in color from pink to blood red depending on its quality. Mining techniques using quicksilver to abstract gold and silver from ore vary, but the mineral was in high demand during the silver and gold rushes in California.

By this time, the Ohlones had already known of this mine for a great deal of time and had, in fact, even brought some of the cinnabar back to the Mission Santa Clara of 1777 to paint the adobe brick.

According to archeologist Mark Hylkema, long before Californians had found the quicksilver mine of use, Indians throughout the western region had traded with the Ohlones for the mineral. "In 1840 the cinnabar trade had broken down because the missions had taken control of the land and the Ohlones were no longer allowed to mine there. One day a group of Walla-Walla Indians dressed in feathers and full regalia came in on horseback with fur traders. They looked fierce and were scaring people as they rode into San Jose to ask why the cinnabar trade had broken down. The cinnabar was quite valuable to Indian tribes who used the powder made from the mineral, stunning vermilion red in color, for body paint. The Walla-Walla *detachment* left without incident."

By the 1870s the Quicksilver Mining Company employed a sizable work force at the New Almaden mine, and two separate communities had sprouted up in the area — Spanishtown and Englishtown. A mine superintendent in the 1920s noted that "Mexicans and their families would not be permitted to take their resident within the sacred precincts of the English camp." As a result of this racial prejudice, each community was self-contained and had its own infirmary, school and church.

Fights ensued over the mine's land for a number of years while California's identity was changing. Eventually the property would have a number of owners and would be mined until the 1950s. Ironically, the quicksilver taken out of the mine would equal more wealth than in all of California's gold mines. Mercury would also be used in one of the next big changes for the Valley, the vacuum tube.

The miner's cemetery at the top of the Almaden Quicksilver Mine ridge
Photo by Sally Richards

Hacienda School in Spanishtown, New Almaden, in September 1908
History San José

35 | CHAPTER TWO

The site of the New Almaden Mine is now part of a county park.
Photo by Sally Richards

(Opposite page) Dane Andrew, who had family on both sides of the Civil War, stands in front of the newly renovated Union Soldier at Redwood City's Union Cemetery.
Photo by Sally Richards

The beautiful ridges surrounding the New Almaden Mine have now become a part of the county park system. Today, the rewards of a long hike will reveal a ghost town that was once home to more than 2,000 people. The ruins of wooden and brick frames where a schoolhouse, churches and modest homes once stood are mostly what's left of the microcosmic quicksilver empire. Further up the trail is a ridge overlooking another beautiful, lush valley. On top of the ghost town's cresting ridge is a cemetery where many of the miners are buried.

A Sobering Mark on California's History

At the height of the Gold Rush, another set of problems came about for the new state. Just as it was celebrating newly found wealth, drunk with new enterprises and delirious with enthusiasm, the state had to sober up to its new responsibilities as a growing power in the country. Many believe the Civil War had no effect whatsoever on Californians, but the state was quickly brought into the politics leading up to the war. California's strong Union sentiment kept Southern loyalists lying low and under suspicion; the state began its own divisions. There were passionate sympathizers on both sides, but California's Electoral College gave its four votes to Lincoln and the state's resources were pledged.

It was at this same time that railroad tycoon Leland Stanford was elected governor by 23,000 votes. One of his first duties in office was to pass stacks of resolutions pledging California's allegiance to the Union. If asked, the many Union supporters would probably not have thought that the South would actually fight the federal government, but on May 11, 1861, all business was suspended in San Francisco and people rallied in the streets in support of the Union.

There are many stories about Confederate loyalists starting skirmishes throughout California, Arizona and New Mexico. Some of California's men who rose to the duty of protecting the Union, such as William T. Sherman and Joseph "Fighting Joe" Hooker, gained notoriety and a place in history for their efforts. In all, 16,000 men from California answered the call to fight for the Union. Those who couldn't fight themselves sent gold to help the cause. California made substantial contributions to the war, and the war may not have been won without its help. Many of the Union soldiers

are buried in GAR (Grand Army of the Republic) cemeteries throughout California.

California gave Lincoln 30,000 majority votes for his second term of office and the state had a deep commitment to the man who wanted to free the south. News of his assassination brought deep sadness for the people of this state who believed in his dream.

Tragic Happenstance Produces a Millionaire

It was at this time when one of the miners who had visited California during the height of the Gold Rush began making news of his own. George Pullman, who had come to California to mine for gold, was an industrious entrepreneur with a knack for making a quick buck. When the get-rich-gold-mining plan fell through, he began selling high-priced supplies to gold miners. After making a $20,000 profit, Pullman returned home to Chicago in 1864.

Pullman, history has it, rode many miles on trains and in much discomfort. People riding the rails in those days could feel every bump through a train's thin wooden seats. On one such journey, the idea occurred to Pullman to build a luxury coach that could be attached to a train — a coach that would rival anything on the railways and which people would pay a great deal of money to ride.

Upon his homecoming, Pullman built the prototype for his sleeper cars, an innovation that would allow America's nobility to move about the country in complete comfort. The biggest obstacle Pullman faced was not building the coach, but building a coach that would have the ability to ride the railways. Although the railways were generally uniform in design, they were not of a standard size.

His problem was solved when John Wilkes Booth pulled the trigger and assassinated President Abraham Lincoln. Mary Todd Lincoln had a tough ride ahead of her, in more ways than one. Not only had she lost her husband in a traumatic murder she had been witness to, but she also had to accompany her husband's body on a painstaking funeral procession on America's railways. To make her journey more bearable, she requested the use of George Pullman's Palace Car, the most luxurious in the line of cars she had seen mentioned in the local press.

The coach, 60 feet long, included an observation room, a bedroom, dining room, toilet compartment with a tub, kitchen, office and storage unit. There was a small space for entertaining in the dining area and the interior was dripping with velvet tassels and fringed drapes and covered in plush materials, rare woods, brass, gold and wall-to-wall carpeting.

When the First Lady requested the Palace Car be added to the President's railroad procession, a world of wealth and opportunity opened up for George Pullman. As the president's rail cortege moved slowly from Washington, D.C., to Illinois, workers simultaneously rebuilt the rails to fit Pullman's coach. Word traveled quickly of Pullman's P.P.C. (Presidents Private Car), as he named the model, and soon it was the rave of high society. By 1867, Pullman's fleet of luxury railcars grew to 48 and his company shipped coaches worldwide as he made hotel cars for railways in Europe, Canada and Mexico.

Had Mary Todd Lincoln not requested Pullman's coach, it's difficult to say whether he would have become a success. Timing had been on his side and he was able to prosper from one of America's most tragic moments. It stands to reason that without that anomaly of happenstance, the Peninsula never would have known one of its richest and most eccentric residents. It was at the celebration for his newly found success that Pullman met Hattie Emily Sanger, a railroad heiress. After a short courtship, George and Hattie were married in Chicago and daughter Harriett was born. Harriett Pullman-Caroland, George Pullman's daughter, would eventually build the largest estate east of the Mississippi, The Carolands, in the young town of Burlingame. It would be more than 20 years before she made her debut in Burlingame, but the city would never be the same again.

A Woman of Great Influence in a City of Little Tolerance

By this time in California's history, there were all types of people coming west for various reasons. Many were motived by greed, and some sought the climate for health reasons, but there is hardly a story more unique than that of Mary Hayes Chynoweth who is now buried in a modest family plot with her family in San Jose.

The Hayes Estate, now a flourishing conference center in South San Jose, is one of those extremely lucky buildings that has withstood the trials of time. The estate, built by Mary, one of the wealthiest women in America at the turn of the 20th century, was a woman whose influence changed San Jose's future. Mary was the matriarch of a powerful family that would bring help civilization to Santa Clara County.

It wasn't long ago when the streets of San

Jose weren't paved and the tallest building in town was only a single story. Arson fires, hold-ups, murder, public hangings and the placement of criminals in stockades for public humiliation were commonplace. At the time, a Mafia-like political machine ruled San Jose.

The Hayes Estate story begins in 1825 in Holland, New York, when Mary Folsom (later Mary Hayes Chynoweth) was born to a freewill Baptist minister and his wife. Mary showed an uncanny ability to heal sick people by soothing them with her hands. One day, at age 27, this unusual woman was doing the dishes when she fell to the floor "like a 100-pound weight had fallen upon her," as recorded in the unpublished memoirs of Elystus L. Hayes, grandson of Mary Hayes Chynoweth.

"When Mary began praying in tongues her father asked who was speaking through her. The voice identified itself as 'The Power.' It told him that Mary would be preparing to do the work of God. She soon began lecturing and giving sermons to the people who visited her home. Although her family was not financially well off, she refused to take money for her services."

Mary and her family moved to Wisconsin where she was married to Anson Hayes, cousin to Rutherford B. Hayes. Their union was cut short when he died of an illness that Mary could only temporarily ward off. They had three children, one of whom died at the age of 4. Her two boys, Everis Anson (E.A.) and Jay Orley (J.O.) Hayes, would go on to college and become attorneys.

The Power became a natural part of life for the Hayes family, often giving advice that proved to be solid. It advised her sons that the property they had bought for $850 would soon be worth $5,000. Shortly thereafter, they negotiated that exact amount from a railroad company. The Power then told the family that vast amounts of ore were buried in Ashland, Michigan. Although surveyors told the brothers that they were "throwing their money down a hole," their mine ended up giving the family a financial freedom they had never known before.

By 1887, both sons had started families of their own, and The Power gave Mary the word to move west. After traveling thousands of miles with her entourage to locate their new home, she found a grove of oaks she had envisioned earlier. She announced she would begin building her estate on that very same property.

Mary had brought many friends and family with her on the quest for a new homestead. At any one time, as many as 30 people were living on the property. Mary added rooms to the original ranch property on the land, and even built four additional houses on the property. It was apparent that a much larger home was needed to accommodate the group. She contracted architect George Page to build the elaborate 50-room Queen Ann Victorian.

The showcase home soon became the axis around which San Jose's social community revolved. Disgusted by the state San Jose was in, her sons decided to wage a political battle against the city's corrupt politicians who had gained control of the *San Jose Evening News*. The brothers formed the Good Government League and supported politicians who were honest. Recognizing the power of the press, the two purchased the local papers, *The San Jose Herald* and the *Mercury*. The papers were later combined to form today's *San Jose Mercury News*.

"When they first bought the papers, San Jose was a pretty rough town," said Daphne Hayes, wife of J.O.'s son, "and getting worse every day. The brothers didn't like the idea of living in such a place, so they proceeded to clean it up. Uncle Everis and father Hayes exposed all the crooks for what they were — they knew where all the bodies were buried."

Mary Hayes Chynoweth
Sally Richards Archive

Both sons openly admitted that they sought the advice of their mother in their personal and political affairs. Mary was quoted as saying in the *San Francisco Examiner* in 1902, "Religion and politics must go hand in hand if the community is to advance along right lights. People should not withdraw themselves into their churches and pray for strength and purity unless at the same time they exert all their influences as citizens to correct the faults in politics."

The brothers continued to prosper in their national-level political careers, and J.O. was later founder and president of the California Prune and Apricot Growers Association, currently known as Sunsweet Growers, Inc.

In 1889, more than two years before the new house was completed, Mary married Thomas Chynoweth, a friend of the family whom Mary had cured of blindness years before. She was 65 years old, he nearly 20 years her junior. Just months before the house was completed, Thomas died of health complications due to epilepsy.

Although advancing in age, Mary did not slow down. She founded the True Life Church, based on The Power's beliefs. Followers would travel from all over the state to witness The Power speaking through Mary during her sermons. When the original church on the Edenvale property burned down (by a careless janitor's cigarette) in 1903, she worked with George Page to design a new church that still stands today (the Unitarian Church on St. James Square).

Mary found religious and metaphysical philosophies fascinating. Some say that Mrs. Leland Stanford, co-founder of Stanford University, and Sarah Winchester, builder of the magnificent Winchester Mystery House, would hold spiritual meetings and have long discussions about life after death. These women were the power brokers of their time and each left a structural legacy for others to

marvel: Stanford University, the Hayes Estate and the Winchester House.

Mary's new house was completed in 1891 and the extended Hayes-Chynoweth family moved in. Eight years later, the estate burned to the ground.

The cause of the fire was never determined. Confident in her beliefs that all would turn out for the best, she wrote in a letter dated September 2, 1899: "The house was needed for some other purpose and we needed the experience else it would not have burned."

Mary began plans to build another home, insisting that George Page design one that would withstand severe fire damage or natural disaster. She worked with Page to design a home that was physically divided into three sections (even in the basement and attic) so that a fire could be stopped before it spread. While much of San Jose was leveled in the 1906 earthquake, the new 60-room Hayes Estate sustained no structural damage, and withstood the 1989 earthquake as well.

In 1905, two months before her new house was completed, Mary died with her family at her bedside. Her last words were, "I have never harmed anyone."

The first Hayes home that burned to the ground
Hayes Family Archives

The current Hayes Conference Center
Photo by Sally Richards

LELAND STANFORD BUILDS A MECCA OF HIGHER EDUCATION

At about this same era in history, another person was making his way from Wisconsin to California. Life's circumstances slowed Leland Stanford down only long enough for him to learn some lessons from the school of hard knocks. A farmer from the beginning, Stanford, the fifth of eight children, began working at the age of 6 cleaning horseradishes for the market. Having grown up in Albany, New York, he then moved to Wisconsin. Rumor has it that he began falling in love with Jane Lathrop while they were still in their teens. But the couple waited to marry until he had an established law firm in Wisconsin.

Soon the self-taught attorney, who knew that knowledge was a powerful ally, had the most extensive legal library in all of Wisconsin. He ran for the office of district attorney on the Whig ticket and was defeated. Then, one day a fire consumed his office and his prized library. The last straw was a stolen cord of wood taken right from the back of his house.

He asked young Jane, "What do we do next?" Her answer was swift, "Go to California!" Five of his brothers had already gone, but since Jane had an obligation to take care of her ailing father, he struck out on his own to the Golden State. He settled and set up a small grocery store near a gold mine that soon failed, and then another near a more successful camp. He finally received word of Jane's father's death, sold the store and went home to comfort his wife. Within a month Jane, who was also fed up with Wisconsin, and Leland decided to go back to California. This time Leland opened a grocery store in Sacramento, the state's capital. He teamed up with others who would soon make up the Big Four (Stanford, Collis Huntington, Mark Hopkins and Charles Crocker).

Stanford, a leader in a small group organizing the Republican Party in California, was the party candidate for state treasurer in 1857 and governor in 1859. He lost both elections, but the party was gaining recognition. Finally, in 1861 the California Republican convention nominated Stanford for governor and he won. Stanford succeeded in holding California in the Union, and strongly and successfully rallied the state's wealthy to contribute substantially to Union victory. Later, Stanford was elected to the U.S. Senate in 1885 and re-elected in 1891.

Stanford's role in building the first transcontinental railroad was undeniably one of his greatest achievements.

Southern Pacific Railroad
Van Court Collection, Redwood City Public Library Local History Room

The majority of San Francisco's businessmen were satisfied with the profits they were making by bringing in goods by sea, and few were tempted by the tremendous undertaking of partnering to build a hazardous rail route that would travel the jagged mountains of the Sierra Nevada. Stanford believed that this project would turn its developers into the kings of the country, but many feared that the project could just as easily turn them into penniless fools. Stanford and his friends took the gamble and formed the Central Pacific Railroad Company to connect with the westward-building Union Pacific.

Before the telegraph was established in San Francisco, news traveled slowly to the West. In 1860 the famous Pony Express was established and recorded its fastest delivery time of seven days and 17 hours when sending the news of Lincoln's inaugural address to San Francisco. The Pony Express, which included showman William F. Cody, "Buffalo Bill," before he made his big break, was too expensive for most people to use at $5 an ounce. California needed another route of transportation, and the railroad fit the bill.

Stanford was elected president of the venture. Congress voted to give the group land grants and bonded loans, but the main monies had to be raised by the men involved. Stanford, Collis Huntington, Mark Hopkins and Charles Crocker risked their wealth and stopped at nothing to meet the Union Pacific at a point as far east as possible. On a clear day on May 10, 1869, Stanford raised a hammer made of Nevada silver and brought it down on a spike of California gold to bury it into a laurel railroad tie. The transcontinental railroad was completed. Gov. Stanford stood before an audience of 3,000 applauding government and railroad officials and track workers in the desert at Promontory, Utah. One of his greatest achievements had just been accomplished.

In 1868, after 18 years of a childless marriage, the Stanfords had a baby boy. It was 1869 when their only child, Leland Jr., celebrated his first birthday. Before he was 2 years old the Stanford family made its first trip across the continent by rail. Feeling the need to make a home for his family, Stanford built a mansion on what is referred to today as Nob Hill. It wasn't until 1876 that the family purchased its first parcel of land on the Peninsula that would become its Palo Alto Stock Farm and later the site of Stanford University.

Photo by Sally Richards

The Stanford family
Redwood City Public Library Local History Room

On The Farm, Stanford used his own theories of blood lines and training that he had learned on the farm of his youth to breed trotters that set 19 world records.

Leland Stanford Jr. found great joy reveling in his parents' adoration and loved their life on the farm. He acquired many animals to keep him company and learned all of the responsibilities of farm life, including how to use its machinery. Stanford had even built a 400-foot miniature railroad for Leland Jr.'s entertainment. At age 15, he was living a good life and wanted for nothing. He studied hard, learned French fluently and developed a deep interest in archaeology.

In 1884, two months shy of his 16th birthday, Jane and Leland were devastated when Leland Jr. developed typhoid fever while on their family vacation in Athens. In hope of curing him, they took him first to Naples, then to Rome and finally to Florence where he would recover briefly and then die. Stanford, who was on his son's deathwatch with Jane the entire time, fell deeply into depression. The morning after his son's death, Stanford went into a sleep that would change the course of history. He dreamed Leland Jr. was alive and next to him, speaking to him. He awoke, and with no forethought of having formed an idea, turned to his wife and said, "The children of California shall be our children."

The couple returned to Palo Alto with their son's body and buried him in the family crypt that recently had been built. The couple thought long and hard about how they would adopt California's children. The natural progression was to build an institution where children could better their lives. Eager to begin work on an institute of higher education that they would dedicate to their son's memory, the Stanfords visited Cornell, Yale, Harvard and MIT. They talked with President Eliot of Harvard and discussed three ideas: a university in Palo Alto, an institution in San Francisco that would combine a lecture hall and a museum, or the possibility of developing a technical school. Asked which of these seemed most desirable, Eliot answered, "a university." Jane Stanford inquired how much the endowment should be, in addition to land and buildings, and he replied, "not less than $5 million." A silence followed. Finally, Mr. Stanford said with a smile, "Well, Jane, we could manage that, couldn't we?" She nodded, knowing a project of such magnitude might ease their overwhelming grief.

They created a university that, from the onset, was untraditional and very unlike the Ivy League schools

A young, 28-year-old architect named Charles Allerton Coolidge took on the tall task of developing the design of the school. In April of 1887 he met with Stanford to go over the preliminary sketches. The Stanfords insisted that the cornerstone be laid on May 14, the anniversary of Leland Jr.'s birth.

It was on that day, a day that would have celebrated Leland Jr.'s 19th birthday, that 300 guests attended the placing of the cornerstone. There was no brass band, but a choir. Stanford called his wife to his side for the ceremony and, though tears streamed down her cheeks the entire time, she held her head high. It was both a solemn and a joyous occasion. The sandstone block and bronze plaque was built into a corner of the first Inner Quad building, near the place where the Memorial Church would eventually stand.

Dr. David Starr Jordan, a renowned ichthyologist, was 40 when the Stanfords lured him from Indiana University to become the president of the university. It was upon opening day on October 1, 1891, that President Jordan gave this address:

Stanford University campus
Photo by Sally Richards

they had been visiting back East. The university was co-educational in a time when most were strictly male; nondenominational when most were associated with a religious organization; avowedly practical, producing "cultured and useful citizens."

"Of all the young men who come to me with letters of introduction from friends in the East, the most helpless are college young men," Stanford once said. As the Stanfords' thoughts matured, their ideas of "practical education" enlarged until they arrived at the concept of producing cultured and useful citizens who were especially prepared for personal success in their chosen professions. "I attach great importance to general literature for the enlargement of the mind and for giving business capacity," Leland said. "I think I have noticed that technically educated boys do not make the most successful businessmen. The imagination needs to be cultivated and developed to assure success in life. A man will never construct anything he cannot conceive. I want particularly that females shall have equal advantages."

The Stanford Mausoleum
located on campus
Photo by Sally Richards

> "Our university has no history to fall back upon. It is hallowed by no tradition; it is hampered by none. Its fingerposts all point forward. I shall not try today to give you our ideal of what a university should be. If our work is successful, our ideas will appear in the daily work of the school. Plain living has ever gone with high thinking. But grace and fitness have an educative power. These long corridors... these waving circles of palms, will have their part in the students' training as surely as the chemical laboratory or the seminary room."

There were said to be anywhere from 10,000 people to half that number depending on who told the story. Leland Jr.'s portrait hung from a sandstone building in the Inner Quadrangle where a makeshift stage had been set up. His mother was one of the last to arrive and was greeted by huge bursts of applause. Years later, when she died under mysterious circumstances in Hawaii, an address she had been too emotional to deliver was found among her papers.

Railroad tycoon and successful politician, Leland Stanford, in failing health, died in his sleep only two years later after realizing his dream of opening the university. His funeral, held in the open air of the university's Inner Quad, nearly marked the end of the university.

Jane Stanford, a full partner with her husband in the founding of the university, was faced with full responsibility of the university that had fallen into financial devastation after her husband's death. Her husband's estate was tied up in probate and several of her advisers urged her to close the university, at least temporarily. After two weeks in mourning, Jane Stanford sent for Jordan and told him she had no intention of closing the university. They planned for endless hours to work out a way to keep the university open.

The Stanford cornerstone celebration on May 15, 1887, a day that would have been Leland Stanford Jr.'s birthday
History San José

Expenses were cut and faculty salaries were reduced. A long-forgotten insurance policy for $10,000 on the life of Leland Stanford tided them over temporarily. Then the probate court granted Jane Stanford an allowance of $10,000 a month from proceeds of the estate, approximately what she was accustomed to spending to run her homes. She reduced her personal staff from 17 to three and her monthly expenses to under $400 (equivalent to a professor's monthly salary) and turned over the rest to Jordan to keep the university open.

In May 1894, just as it seemed the university stood a chance of surviving, the estate was tied up indefinitely by a federal government claim of $15 million growing out of construction loans to the Central Pacific Railroad. The loans were not yet due, but the government sought to establish stockholder liability. Jane traveled to Washington, D.C., and asked President Cleveland for prompt action by the courts.

More than two years later, the Supreme Court denied the federal government's claims against the Stanford estate. Although it was raining that afternoon at Stanford, students and staff celebrated long into the night. Jane's estate was released from probate in 1898 and in 1899, after selling her railroad holdings, Jane Stanford endowed more than $11 million to Stanford University's trustees. What Jordan termed as "six pretty long years" had come to an end. An incredibly strong-willed woman, Jane Stanford would not watch the legacy of her son die once again. What Jane was unable to do for her son, she did for his university. Leland Stanford Jr. University would become a flourishing institution of higher education that would change the course of the world.

Change Continued

People continued to flood into the Valley and change every aspect of life as the indigenous people and settlers knew it. Cultures from all over the world were introduced to California and the state was melding them into one of the most diverse populations in America. It's often said that where a community's population is growing substantially, and technology is needed to supply greater amounts of goods, is where technology grows the fastest. So ends the midmorning of Santa Clara Valley as it learns hard lessons, gathers its resources from the most gifted citizens and heads swiftly toward noon's innovations.

Author Bayard Taylor first visited California in 1849, and when he returned 10 years later he wrote the following in his book, *New Pictures from California:*

> "...As we drove along it, I looked in vain for the open plain, covered with its giant growth of wild mustard. The town (San Jose) now lies imbedded in orchards, over whose low level green rise the majestic forms of the sycamores, which mark the course of the stream. ...the valley strikingly reminded me of the Plain of Damascus.... But in place of the snowy minarets, and flat oriental domes there were red brick masses, mills, and clumsy spires, which (the last) seemed not only occidental, but accidental, so little had they to do with architectural rules.
>
> The valley in which it (San Jose) lies is one of the most favored spots in the world.... When the great ranches are properly subdivided, as they will be in time, and thousands live where units are no living, there will be more no desirable place of residence anywhere on the Pacific Coast."

Photo by Sally Richards

Photo by Sally Richards

"To see a world in a grain of sand and heaven in a a wildflower..."

— *William Blake*

CHAPTER THREE

Noon is the longest period of the day; it spans from when midmorning becomes warmer and lasts until the sun begins to dip into afternoon. Noon in the Valley of the Heart's Delight was a period of time between the very late 1800s to the end of the 1960s, a relative span of minutes in relation to what came before and what would follow. During this time, people learned to rely on the luxuries of newly found automations but lived in the anxious moment of tomorrow's discoveries.

As the Valley's people stood on the edge of a new era, changes were taking place that would affect the way the world's people lived and worked. The birth of technology was occurring right in the Valley and the free-thinking people who made these discoveries would set the pace for the next 100 years. During this period, generations and traditions were born and died, and the Valley's residents, some holding on very tightly to their old ways, were brought into the next span of progress. Some are still feeling the growing pains that threaten to change the very way they live; others revel in a modern world, unaware of what has come before.

Growing the Valley

Early residents found that cattle and horses bred well in the temperate climate of the Valley. Relatively speaking, the cattle yards held their ground in the Valley for only a short time. The Valley's soil seemed destined for some grander purpose and yielded well to the changes from natural cultivation aided by the Ohlones' controlled burnings to the pioneers' planting and harvesting of the rich, tilled earth. Miles and miles of golden grain swayed in the gentle winds under the rising sun of the Valley's noon. Men with scythes and thrashers harvested the grains grown by traditional farmers.

The Valley's early harvesters
California History Center

The Gold Rush had brought a flood of immigrants to California, many of whom traveled west from the Sierra foothills into the Valley after their mining dreams went bust. Many of the Valley's new residents came from the warmer regions of Europe, where climates were very near to that of their new home. Recognizing a good opportunity, they began planting trees indigenous to their homelands; their fruit orchards would soon produce some of the finest, sweetest crops ever.

These young, thriving farms and orchards provided work for many Chinese immigrants who had been forced out of mining and railroad jobs as a result of anti-Chinese legislation during the mid-1800s. Soon Chinese communities were developing throughout the Valley, situated nearby farms and orchards in Mayfield, Mountain View, Campbell and Cupertino. But the largest Chinese population resided in San Jose. The first Chinatown, located at Market and San Fernando streets, had all the amenities of a Western settlement, including grocers and barbers, lodging houses, restaurants and even musicians.

A fire in 1870 destroyed the settlement, and though deemed accidental, these "accidents" were not uncommon in Chinatowns throughout California. A new Chinatown quickly emerged from the ashes in San Jose, but this, too, was destroyed by a second accidental fire in 1887. Following the second fire, John Heinlen, a retired farmer and a businessman with an affinity for the Chinese, purchased land for a new Chinatown, which was named Heinlenville in his honor. It was completed in 1888 and was surrounded by a high barbed-wire fence. The gates into the village were locked each night to protect Heinlen's property and the Chinese families who lived there. Though poor migrant workers, the residents of Heinlenville pooled their money to build a temple, Ng Shing Gung, which became their center for worship and community gatherings.

As the Chinese community grew in the 19th century, its younger generation began to integrate into the

A fire in 1887 destroyed San Jose's Chinatown.
History San José

Japanese-Americans gathered at the Buddhist Temple in San Jose for a funeral.
History San José

(Far left)
Heinlenville's temple,
Ng Shing Gung, c. 1900
History San José

A Santa Clara cannery at the turn of the century
California History Center

community outside Heinlenville's gates. After John Heinlein died, his land was sold to the city of San Jose. Eventually Heinlenville was torn down and the land quickly redeveloped. Though the original Ng Shing Gung temple was dismantled, during the 1990s the Chinese Historical and Cultural Project replicated the temple and donated it to the San Jose Historical Museum, now History San José.

The arrival of Chinese immigrants in the Valley was followed by an influx of Japanese immigrants just before the turn of the century. Many Japanese men replaced Chinese workers who were leaving the orchards and fields over labor disputes with white landowners. A labor shortage in the early 1900s, a result of the Chinese Exclusion Act, allowed the Japanese to strike for higher wages and eventually gain rights as farm tenants.

The first Japanese immigrants formed Nihonmachi, or Japantown, a predominately male bachelor society. Over time, more settled communities of Japanese workers and their families developed in the valley. These farm clusters were critical to the success of Japanese farmers, who learned to share tools, resources, labor and even the marketing of their crops.

The ethnic solidarity of both the Chinese and Japanese communities provided protection from and resistance to the anti-Chinese and anti-Japanese sentiments of white landowners in the region. These segregated communities offered their citizens all the necessary resources — schools, businesses, recreation and fellowship — diminishing the need of Chinese and Japanese immigrants to integrate into the larger community.

One by one the grist mills and livestock ranches disappeared and were replaced by fruit processing plants that shipped Santa Clara Valley's fruit products nationwide by boat and train. The lumber industry began fading and many of the barges and boats that once carried lumber from Alviso and other peninsular harbors were now shipping fruit. The many immigrants who had come in search of riches found gold's prosperous hues in the deep colors of the fertile earth and the round, robust apricots, prunes, pears and almonds it yielded. The noon's sun brought out the vibrant reds and purples of cherries and grapes that blanketed the Valley. The seasonal crops swiftly grew into a powerful world currency that would harvest an agricultural empire.

By the 1890s electricity had already been discovered, but many thought it might be better to apply natural resources to generate it. There was a great buzz caused by entrepreneurs and engineers, primarily electrical engineering professors at Stanford, to apply the lessons learned by the hydraulic mining techniques to harness energy for the modern-day Bay Area. It wasn't long before local power companies began building hydraulic plants in the foothills of the Sierras. This would be one of the first cooperative joint ventures between Stanford and industry that continues to this day.

Today's abandoned Alviso Station
Photo by Sally Richards

Grace on her childhood orchard home
Grace Kennerson

Grace Kennerson today
Photo by Sally Richards

(Far right) Brokaw Road, 1913
History San José

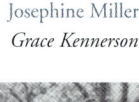

Josephine Miller
Grace Kennerson

By the turn of the century the Valley's floor was carpeted with seemingly enough fruit to feed the world. Thousands of acres of the sweet, plump fruit were being boxed and shipped across the globe. Many farming dynasties were made at the time and their families would manage to avoid change for many decades.

Memories of a Time Gone By

Many Valleyites who were born during the time of the Valley's noon continue to call Santa Clara County home in the new millennium. One such woman is Grace Kennerson, who now lives in the Willow Glen area of San Jose. Her grandparents, the Richard family, came from the South of France. Her father was French and her mother was Irish. Her parents had their hands full managing a farm and raising their household of three children.

"When my mother and father were married, they moved into an old adobe house down by the creek on 237 and First Street. It was so old it was built by the Indians," said Kennerson. "My father thought it was getting kind of run down, so they built another house closer to the road. The house didn't have a furnace — it had a fireplace in the front room, a big old stove in the hall and a sun porch with lots of windows.

"Our farm in Alviso had a lovely breeze that came in from the bay, so it was always cooler than downtown," said Kennerson of her time on the family's orchard. "Every year the yacht club in Alviso would open. The harbor was always busy with men working on boats. There was the cannery, and it was busy too. We never sold our pears there, we sold them to a place in Sunnyvale called Schuckl. Some of the pears I see in the stores today... we would have thrown out. I used to help sort them. I also did a lot of things in the house. My mother would, once in a while, go outside with the men, and she'd say that some of them would be singing while they worked, not quite like it is now. My dad used to ship the pears in boxes and put them on the train that came to Alviso. We also had another ranch over on Mt. View Road — 165 acres — and we rented that out, and the people who leased it had a dairy on it. We used to get our milk from their dairy farm every other day.

"The orchards were beautiful... we had pears and apple trees — everything would be in bloom in the spring," Kennerson said, reveling in her sweetly perfumed memories. "All the way to First Street and into Saratoga ... it was solid blossoms. And to smell them! Everyone would drive all around the orchards," said Kennerson about the people who would come from San Francisco and beyond to see the thousands of acres of blossoms on the weekends. "The whole Valley smelled really sweet, especially the apple blossoms. We would drive down First Street on Sunday and see everyone we knew. There were dress shops and all kinds of stores, but downtown was real small. And the dress shops always had someone there to wait on you so you didn't have to go through all the clothes like you do now.

"We used to go to church on Sundays to St. John the Baptist church in Milpitas every Sunday at 8:30 in the morning. Afterward, we'd drive to the show at the California [the Fox] on First Street. They don't know what they're going to do with it now, but they want to save it. I just don't know," Kennerson said of the movement in San Jose to save some of the older buildings in downtown. "I saw *Dracula* there with Bela Lugosi — that was really scary. The California used to have a stage show, mostly chorus dancing and an orchestra for about a half-hour, and then they'd have the movie. Clara Bow was a big star then. I miss going down First Street and seeing everybody I knew, everybody I went to school with. I miss it not being as crowded as it is today, and the slowness of it all. Everything today seems so fast. Everybody's in such a hurry."

Kennerson and her sister, Josephine Miller, graduated from Notre Dame High School, and Grace went on to attend San Jose State University where she studied English and home economics. She then met her husband, from San Francisco, who would later manage The State Theater in San Jose. The young couple married in 1936.

"My brother attended Santa Clara University and then graduated from Davis. By that time, my father had been killed in an accident when he was driving back from duck hunting, it was raining so hard and his car slipped. My brother and my mother ran the farm for a few years, but my brother got married and didn't want to run the ranch anymore. He got a job in town and my mother was running the ranch by herself. It was hard for her."

Meanwhile, Kennerson and her husband Bert were transferred by the entertainment company that employed him to Pasadena where they mixed with Hollywood's studio royalty. The couple did return to San Jose and bought their house in Willow Glen.

"We didn't have to sell the ranch," said Kennerson of her family's sale of all of their ranches, including another large parcel on Gish Road. "My mother was running the ranch by herself, but it got to be too much for her, so she retired about 1940 and bought a house down the street from me."

Grace and Bert went on to have three children of their own. As life became complicated by World War II, Valley residents helped out with the war efforts in their own way. "There were a lot of war pictures that came to the theater," said Kennerson of the theater her husband managed. "All the theater managers were supposed to sell war bonds during the war and whoever sold a million dollars' worth — and that was a lot of money back then — went to Washington, D.C., to be honored. They were called the Honored One-Hundred — my husband was one of those men."

SUNNYVALE COMES OF AGE

Economies and industries were changing quickly for sleepy Bay Area cities that were only just beginning to blossom. Charles Bocks is one example of a man light on his feet and able to shift well when wheat became a dying business proposition and orchards came of age. He had been a grain farmer in the 1850s and would go on to establish more than 55 acres of cherry orchard located on Hollenbeck Avenue in Sunnyvale. At the time, his crop was believed to be the largest cherry orchard in the world.

Photo by Sally Richards

Theodore Roosevelt when he visited the Peninsula garnering support for his party
Redwood City Public Library Local History Room

Sunnyvale's entrance at a time when agriculture still ruled
California History Center

Yvonne and her mother, Rose
Yvonne Olson Jacobson

Yvonne Olson Jacobson today
Photo by Sally Richards

Yvonne Olson Jacobson, author of *Passing Farms: Enduring Values,* a book known as the Bible of the fruit-growing industry in Santa Clara County, is one of the descendents of the longest-enduring fruit tree growers in the Valley. Her grandfather, Carl Johan Olson, came to the United States from Sweden in 1880, and traveled back and forth across the United States several times before he landed in San Francisco in 1898 and married Hannah Louise Merck, also a Swedish immigrant. While they were looking for a place to build a home, Jacobson's grandfather, who worked in the chemical industry, fell ill due to fumes in the factory. His doctor suggested they move to a location where there was plenty of fresh air. Both had worked and saved their money before marrying, so they were able to purchase five acres of land in Sunnyvale at the turn of the century. In 1899 Jacobson's father, Ruel Charles Olson, was born and the seedlings of the Olson dynasty in Sunnyvale were planted.

It wasn't long before the good economy and mild climate were drawing a constant stream of people from all points of the globe. El Camino Real, a trail used by the missionaries, had become a major thoroughfare between San Jose and San Francisco and it ran through Sunnyvale. Wagons and later cars would eventually turn the dirt road into a nonstop ribbon of traffic recognizable from any of the Valley's highest peaks.

In 1906, just before the earthquake, there was a land boom in the Valley and the young Olson family was offered a great deal more than they had originally paid for their land. All they had to do was believe that they would find their dream elsewhere, sell their property and relocate. "They couldn't bypass the money they were offered," said Jacobson. "In fact, they began to realize that they were right in the middle of town and this would only be an increasing problem. They agreed to buy outside of town where it was cheaper and they were able to buy more land in another place."

The Olsons kept their house and were in the process of moving it to the property on the corner of McKinley and Taaffe. The home had been placed on rollers to be transported to a foundation on their new property. At 5 a.m. on April 18, when the earth shook, the house began to move on its rollers. "My grandfather went down to the basement area and tried to hold the house in place. Fortunately there wasn't much damage done. They were very lucky," Jacobson said of the 1906 quake.

"Most farmers regard their land as their most important asset," said Jacobson of the strong ties that bind farmers to their land. "Some people have the good fortune of being so far away from cities that urban sprawl never becomes an issue. But, there are so many examples in this county, and the whole state, where there are population explosions. I spoke to farmers in 1979 and onward, and they simply said that, when it comes down to it, they can take the money they make from selling their land and for every acre they sell, they can easily buy three or four acres in another part of California. Most consider themselves ahead, but it's not always possible. This is only possible in an up market when there is a lot of cash around and the pressure to build more homes is on the rise."

Jacobson, her brother, Charles, and sister, Jeanette, lived on the Olson farm with their parents, Ruel Charles and Rose Zamar Olson, for their entire childhood. "That's the property where all three of us were raised. We grew up there, we played there. We knew every inch of the place, all the nooks and crannies, all the places to hide, all the trees. I climbed those trees when I was a child... I loved to be in those orchard trees."

The Olsons lived on the edge of the city of Sunnyvale. The farm, situated near El Camino Real and Mathilda, was one of the last orchards in the area; most of what remained in the Valley was south through Saratoga and Cupertino.

"We grew apricots, cherries and prunes to hedge against the failure of any one crop," said Jacobson about the lengths growers had to take to protect themselves against hardship. "Often the weather conspires to destroy

The historic Olson water tower
Photo by Sally Richards

certain crops at a certain time when the fruit is at its peak. It wasn't unusual to have cherries ruined by very strong rainstorms. You could also have some kind of disease destroying a crop of the apricot trees. In August of 1918 there was a disaster — all the prunes in the valley had been picked and were on the ground drying when there was a tremendous rainstorm. It ruined the entire prune crop. There were soldiers stationed at the Fremont Camp in Menlo Park and they were brought in to try and save them by stacking the trays and getting them out of the rain. But it was too late and it was almost a total write off."

The end of the 20th century was the death knell for many orchard trees in the Santa Clara Valley. One of the great tragedies marking the end of the orchard era was the swift destruction of the fruit trees next to Hewlett-Packard in Cupertino. The orchard saw its last season, its trees ripped out of the ground with cherries still on them. The loud cracking of the aged cherry wood as it met the bulldozer's metal is a sound old-timers are far too familiar with.

Another landmark locals notice missing is the Olson's cherry stand once located at Mathilda and El Camino Real, a place that had become a traditional stop for generations. The wooden building is slotted for demolition with future reconstruction in another nearby location. "She loved working with the fruit," said Jacobson of her mother Rose who ran the fruit stand. "As she would pack a basket of cherries, you'd see her pop one in her mouth. It's not that she didn't like what she was selling — she was her own best advertisement.

"Both of my parents passed away," Jacobson said about the sale of a 10-acre property to a developer in 1988. "The IRS stepped in to claim its share of the inheritance and eventually one orchard had to be sold. This wasn't the first time we had watched our past being ripped up. It happened in the 1960s when two of our other orchards were demolished in an eminent domain ruling."

Jacobson, a woman of great tradition, has managed to hold on to the remaining three acres of Olson property with some of the last cherry trees in Sunnyvale. The remaining Olson property was divided among the three Olson children. Jacobson's sister and brother took over a 16-acre piece of property and recently leased it to a developer.

Jacobson was recently working on a story about the history of the Olson family for a workbook intended for the local school district. One of the days she was working on the project happened to be the 100th anniversary of her father's birth.

"The tremendous irony of the whole thing was that I was asked to write a history of the family for the grade schools," said Jacobson of a day that she will remember her entire life. "On the day I was writing it, which happened to be September 7, 1999, it was exactly 100 years from my father's birth — it was also the day they were pulling the trees out. It takes them less than two minutes to pull out one of those trees, but it took about 100 years of cultivation to get them all in place. He always said he wanted the Olson orchard tradition to last 100 years and it did. It was one of those historic ironies. I was out there watching... it was very hard. At the same time, it had been a long time coming and there was a lot of preparation and knowing it was going to happen. But when you see it... the transformation, it's an incredible loss and change. It's certainly a loss of one's past, in one sense, but of course you carry with it a tremendous amount of memories that live with each of us for rest of our lives. There is a nostalgia for what has gone — at the same time you have to move on in your life, you have to make peace with the past and adapt to the present... otherwise it's unrealistic."

The clearing of the 16-acre orchard made way for a development of 300 apartments and more than

One can hardly go a block in Silicon Valley without seeing the effects of urban sprawl.
Photo by Sally Richards

Del Monte Foods

My son is a professor at Stanford teaching environmental science — in a sense he's still on the side of the earth — and he has two little children who I doubt very much will grow up to be farmers."

For generations children and teen-agers were part of a standing tradition that occurred every summer when the fruits of the Valley were bursting with flavor. This was a time when kids would set aside their bikes and toys and turn nearly all of their efforts toward picking the millions of buckets of sticky, sweet fruit from the trees that towered over them, or scooping the fruit off of the ant-covered ground. Summer school would have been a welcomed respite for these kids who toiled under the noon's sun. Not until school reconvened were their fruit-stained hands able to rest. Carol Beddo, San Jose author, has mixed remembrances of picking prunes on Black Road in Los Gatos during the 1950s:

60,000 square feet of shopping center to accommodate Silicon Valley's growing population.

Jacobson's family set a standard for the generations that would follow them. "My father had a great sense of the history of the place, and my mother did too. She had worked all her life... very, very hard. Their life was one of hard work. Farm life can't be romanticized because the difficulties are one of the reasons people leave farming. You have to enjoy being outdoors and dealing with the frustrations of those parts of your life that you can't control. It's all involved with nature in a very elementary way. My mother's purpose in working so hard was to pass all this down to her children and her grandchildren, so there was a sense of satisfaction that she had done that, and done it extremely well. Her hard work had yielded an asset that was more than that, it was a way of life that could be passed on and continued.

"Life simply changes and there are very few people today who can predict what their grandchildren will be doing," said Jacobson of the ability to pass on an agricultural business. "It was a wonderful thing to carry on a tradition, it just happened to work in my family for four generations spanning 100 years. Things have changed, and life changes. My father never wanted to leave this area. That's why we're here, even though the farm is gone.

The Prune Pickers

Summer's true end is to be up at first light, heading out to the orchard, empty bucket in each hand. Squat down, pick up purple prunes lying among dirt clods. Both hands work together. Quick, quick. Hands full. Hurry up. Faster. Each hand holds three, four dusty prunes.

This is summer's end; school will not start until the prunes are picked, and we are expected to pick them. All students expected, prune pickers all. Wanting to has nothing to do with it. Be there, or else.

Drop six, eight prunes at a time into buckets. Plop! Hard prunes thunder against empty bucket bottom. Start each bucket with hard prunes on bottom, finish with soft on top. Keep two buckets going. Gather more. Drop.

Gather. Steady, repetitive, still in squat. No standing. Never sit, never kneel. Move like a duck over dirt, around tree trunk.

Fill the buckets. Hurry, hurry. Ignore leg pain. Ignore back pain. Forget neck and shoulders. Stand, at last, with full, heavy bucket in each hand pulling out shoulder kinks. Long, deliberate steps stretch legs heading toward stacked mountain of old, empty prune boxes. Boss chalks my initials on weathered, wooden box end. Hurry, hurry. One by one, each filled box adds 15 cents to my tally. Five, six buckets fill each box. First light to sundown, each bucket takes longer than the last.

Temptation calls. Boss won't notice a few dirt clods. Anyway, a few get in there all by themselves. What's a few more? Everybody does it. Right?

Think again. No glory in this rebellion; summer's shame is in getting caught. Do we sometimes sin to survive? Is it ever okay? Will it ever be?

Mid-morning sun gets warm. Now the whole day will steadily heat up. No cooling off on these days. No cooling until deep in the night when prune pickers sleep close to wide-open screened windows. My aching body is covered with a crisp, white line-dried cotton sheet.

Hotter and hotter all day long. Try to pick in the shade. But someone has to pick in the sun. So make a deal. Take turns. Share a tree. Picking partners trade shade on this tree, sun on the next.

Stop for lunch. Walk down the cool canyon, where a spring trickles its last before winter rains arrive. Sweat tickles face, neck and back, sets dry with gray dust. Crusted fingernails show against white bread. Eat, drink slowly. Relax, rest. Redwoods close off hot sun, tenting us in dry, cool shade, tempting us to doze.

Boss calls us back rudely and too soon — we don't have all day! Okay, okay. We have a half-day. Boss is someone we don't have to like. And we don't.

Early drop prunes dry black, wrinkled in the sun. Pick them anyway — they are prunes, too. Eat one now and then; fruit leather encloses hard, sweet pit, summer's true taste. In shade, early drops ripen, go soft, attract bees and ants, ooze hot, sticky sugar that absorbs dust in dark, dry patches. This is how summer smells, hot sweet dirt caught in my throat.

Ants embroider black lines across the orchard floor, silent crisscrossed trails connecting, crossing dirt clod mountains and valleys. Each wiggling walking forward creature creates the whole. Their design features black ovals — ripe, oozing prunes covered solid with working black ants. They work for sugar. Black prunes everywhere I pick. Prune harvest includes these victims in buckets, in boxes, destined for extermination at the dehydrator. Pure chance. A crowd of ants on one ripe prune, multiplied by thousands, never to rejoin their black line, never to return to the nest. How do survivors manage? Can their broken lines reconnect? Should I care? Just get back to picking before Boss tells me to.

Boss does not expect us to like picking prunes. And we do not. But liking to, just like wanting to, has nothing to do with it. The more of us who turn out, the sooner we finish.

Boys always throw prunes. Boss always stops them. They mainly throw at each other. They also throw at girls. Boys wait till boss is busy, not looking, to start silent prune missile wars, silent until laughter erupts and boss yells at them — quit it! You could put somebody's eye out. Get back to picking. And I mean now.

School could start on time, a few days late, or a week or two late, depending. Prunes are the valley's calendar; prune picking, its test of character.

— *Carol Beddo*

Del Monte Foods

Changing Times

The turn of the century was both blissful and devastating for the Bay Area. The Valley's economy was booming and entrepreneurs flooded the area with new inventions of industry and service. Miles and miles of sweet-smelling orchards, at one time 125,000 acres of fruit trees, covered the Valley's floor, standing tall and robust against the stunning backdrop of clear blue sky. The bountiful tree era in the sun-drenched Valley of the Heart's Delight propagated the belief that a boundless future was possible.

Farmers had discovered what grew best in the Valley and what industries the area would sustain. Resources such as wood, brick and water were rising in cost. By the turn of the century, drinking water had reached $2 a gallon in San Francisco, but soon dams, reservoirs and lakes would be built to support the Bay Area's growing population.

Then, on April 18, 1906, at 5:13 a.m., under the rich soil that produced perfect rows of fruit-bearing trees, the earth's plates shifted. Buildings, quickly constructed to meet housing and retail space demands, crumbled into piles of splintered and crushed materials. In less than a minute, a prosperous area was beaten down as if by the fist of a violent god. Buildings tumbled and fires leapt from house to store with no means to stop them since many of the water mains had been damaged by the quake. Up and down the Peninsula havoc reigned. No matter a person's station in life, no amount of money or education spared or softened the blow from those 60 seconds when thousands of people thought the end was near.

As with the 1989 Loma Prieta earthquake (7.0), damage was reported all the way to Hollister on the coast and into the North and East Bay. The 1906 earthquake was a rocking 7.8 and bled all areas of the basic resources of fresh water, food, shelter and medicine. The 1906 event is said to have been 16 times stronger than its younger sister, Loma Prieta.

Although worldwide attention focused on the catastrophe in San Francisco, which was destroyed by fire following the earthquake, communities south of the city also lost many lives, homes and businesses. At Stanford University, the library and gymnasium both collapsed and the Memorial Church was damaged. In many of the Valley's orchards, the fruit trees were uprooted and fell against one another. Even the already dead were not spared from quake damage — in Colma's marble orchards, tombstones, hand-turned wrought iron fencing and mausoleums shook enough to require months of cleanup. The quake caused irreparable damage to now rare turn-of-the-century handcrafted West Coast funerary art.

The San Jose Post Office
after the earthquake
History San José

In and around San Jose, most of the fatalities were caused by the collapse of the facilities at Agnews State Hospital for the Insane. Over 100 patients and employees were killed and many others were injured. After the earthquake, the dazed inmates wandered the hospital grounds and surrounding streets unsupervised. Dangerous patients were tied to trees until they could be transported to another facility in Stockton, while others were gathered under palm trees on the hospital's lawns.

Sand Dreams & Silicon Orchards | 60

What follows is an excerpt from the front page of San Jose's *Mercury-Herald* on the day of the earthquake. Printed just a few hours after the disaster, the type was hand set because the paper's machines were badly damaged. Later newspaper reports provided more accurate accounts of the lives lost and property destroyed by the earthquake.

"In San Jose evidences of ruin are only too apparent. The business section is naerly destroyed. What structures remain standing have fractured walls and must be torn down. The dead in the city number sixteen. At Agnews, where the buildings were completely destroyed, the dead, patients and attendants, number about 125. The work of rescuing the living and their silent companions from the ruins has continued unceasing all day, but many are still buried in masses of brick and broken lumber. Hundreds of deputies, hurriedly sworn in by Sheriff Ross and taken to the scene, guard the uninjured insane, who are clamoring on the lawns in an ectasy of fright and fear.

"In San Jose the hospitals that remain standing are caring for great numbers of injured, many of which are hurt mortally. Immediately after the fearful concussion of the 'quake fires broke out in several quarters of the city. The El Monte lodging house on Locust street took fire immediately after the collapse. Seven people, two entire families, were roasted to death. Twenty persons were imprisoned in the ruins of the Vendome hotel annex, but all were reported living with the exception of one. Two persons were killed in a collapsed building on Market street, just north of Santa Clara. One fireman was killed while at his work of rescue. One woman was killed near Santa Clara by the collapse of a water tank on her house. Another woman was found dead in a house on Devine street. Two patients at the county hospital were killed."

— *Mercury-Herald,* April 18, 1906

After the earthquake, the populations of Bay Area cities began to grow with the throngs who had fled The City by the Bay. With the commitment of San Francisco's

Agnews Sanitarium following the 1906 earthquake
History San José

After the earthquake, National Guardsmen set up a temporary camp in St. James Park to keep order in San Jose.
History San José

The Joshua Hendy Plant and the men who worked there
California History Center

One of Sunnyvale's first real estate offices
California History Center

expatriates, new companies were built and brick and lumber formed sturdier places to live and work. The people of this age were of a resilient stock, fearless women and men who braved a plethora of dangers to come west; there was no way that a disaster lasting less than a minute was going to put a damper on their future plans. This was, after all, California, a state with the motto "Eureka!" from the Greek word *heureka*, meaning "I have found." Whatever they found, it's certain that people kept finding it, because the state grew and people continued to prosper.

Near the sun's highest point of the Valley's noon, many large companies were established in the Bay Area. Joshua Hendy Iron Works, one of Sunnyvale's largest companies, had originally been founded in San Francisco in 1856 but after its plant was destroyed in the Great Quake, it found sounder ground in Sunnyvale near Evelyn Avenue. Joshua Hendy Iron Works assumed critical roles in the war efforts of both World War I and World War II. Westinghouse purchased the Iron Works in 1947 and continued to build a wide variety of heavy industrial and defense products, becoming the corporation's Marine Division in 1965. The plant was sold by Westinghouse in 1996 and is now operated as Northrop Grumman Marine Systems.

CITIES OF A GRANDER SCALE

Many exciting things began happening after the Great Quake. Colorful tycoons began building legends that would live on to this day. The common folk of the day watched in amazement and disbelief as the soap opera-like stories unfolded with much fanfare, most of it reported in the daily papers — especially in the gossip columns. The Bay Area had a common goal to ready itself for the 1915 Panama-Pacific Exposition, an event that encouraged the construction of even grander estates and cities. The preparation for the expo reintroduced a sense of wonderment that the Valley's and San Francisco's residents seemed to have lost after the quake.

As the dust from the earthquake settled and new life bloomed, another time had arrived, a time when creative entrepreneurs rose from the smoldering ruins like phoenixes. It was a period when millionaires were made by repairing and rebuilding the damage caused by the quake. Brick and lumber kings would grow richer from restructuring the Bay Area, and once ordinary people with determination and carpentry and mason skills were now in great demand.

Retail stores, banks, real estate offices and service companies were opening all over the Valley to meet the needs of its growing population. Manufacturing companies were getting a hold in the local markets and the working

population flourished. People were excited about the good times, and even some of the wealthy lived well beyond their means, believing in their hearts that they would not only recoup their loses, but also achieve mega-wealth.

San Franciscans of "old" money, reaching back a mere generation, realized long before the quake that to truly "arrive" in society, they needed a second "country home" in the hills of the Peninsula or the East Bay where the pace of life was slower. Far from the blue-collar work of the farms, canneries and mining in the South Bay, the Peninsula offered country clubs and train stations where private rail cars would arrive and disembark daily. Grandly dressed women and men left the Peninsula to enjoy fine meals and live entertainment in The City. People traveling in elite circles left San Francisco to arrive at the Burlingame and Belmont stations where they would be whisked away for entertainment at some of the most posh estates in the country and to participate in polo games and fox hunts. Bankers and railroad heirs hired the trendiest architects to build the most elaborate estates, and the "country homes" grew larger and grander with every new millionaire made.

Great mansions seemed to grow from the very earth as the wealthy tried to outdo one another. Wealthy Peninsulites would soon need to entertain the royalty and celebrities the Panama-Pacific Exposition promised to draw. The fabulously rich became eccentric icons whose only limitation of fantastical opulence was the amount of money they could draw from their accounts. Times were extreme in the Bay Area — while those who were rich threw their money around with great extravagance to impress others, those who failed walked into the bay and drowned like William Chapman Ralston. It was a place of stunningly impressive beginnings and horrifically tragic endings.

THE PENINSULA'S DRAMA

One such story of wealth and tragedy is of a woman who went from the grandest scale of economy to the drudgery of commonality. Her name was Harriett Pullman-Carolan, and like many women of wealth from that era, she had inherited a great deal of money from her father. As with many of these heiresses of silver and railroad fortunes, men were easily accessible and quite persistent.

Photo by Sally Richards

Harriett's father, Chicago railroad king George Pullman, was made wealthy by the invention of the Pullman private coach, an elite private rail car in great demand by the world's aristocrats.

Harriett's home, once the queen of the hills, was dubbed the Carolands estate. The home, currently covered in scaffolding, shows its battered beauty and offensive neglect as testimony to the many years during which interim owners were able to keep the wrecking ball at bay. Located near the Black Mountain Highway at Highway 280, the massive 112-room mansion, with nearly 1 million square feet of space, hides behind locked gates in an exclusive Hillsborough neighborhood. The grand structure looks like an abandoned château plucked from 17th-century France and accidentally dropped into a neighborhood of substantially smaller, suburban homes.

Happenstance plays a large part in the many twists and turns in the saga of Carolands. The story began in 1869, a year that held two blessings for George Pullman and his young bride, Hattie. It was the year of Harriett Pullman's birth and the driving of the golden spike linking the Union Pacific and Central Pacific railroads that allowed the Pullman coaches to travel from the Atlantic to the shores of the Pacific.

The Atlantic/Pacific railroad line is what eventually brought together their daughter, Harriett, and Francis Carolan, the man who, much to her father's chagrin, would become her husband. Supposedly, Hattie, a hypochondriac, was searching the West Coast for a cure for one of her many various ailments. Somewhere along the train journey, Harriett met and fell in love with the dapper Francis "Frank" Carolan.

When Harriett Pullman, heir to the Pullman fortune, set her sights on Frank Carolan, former manager of a failed family-owned San Francisco hardware store, her parents were far from pleased. Carolan, who graduated from Cornell University in 1882, was described as a handsome, debonair gentleman and a favorite of the San Francisco bachelor set. Harriett was educated at a fine New York finishing school and spent much of her time abroad learning about the finer things in life. The couple, different in many ways, fell in love.

On June 8, 1892, the day of their wedding, her father gave the couple $35,000 to build a new home and $25,000 in Pullman-Palace Car stock. As time went on, their stock grew and the couple made wise investments.

It wasn't long before the couple moved from their modest San Francisco residence to Burlingame and into a home on a 30-acre property they named the Crossways Farm, an area now located between Burlingame and Oakgrove avenues. San Francisco architect Willis Polk's plans for the property included several formal gardens, a bleaching green, stone pergolas, tennis courts, hot and green houses and dog kennels. It is believed that the garden layout was the design of Polk's friend Bruce Porter. He and Porter had completed work on another such project nearby known as Filoli (*Fi*ght, *lo*ve, *li*fe).

The couple's neighbor, the Burlingame Country Club, was unable to finance a polo field on its own. Anxious to attract high society's approval, Carolan erected elegant buildings on his property where club members could watch horse racing and polo games. He even built bleachers where hundreds of cheering spectators could be seated.

John Bromfield remembers a polo game of 1913: "The San Mateo Slashers had beat the Hawaiian team, ten-and-a-half to five-and-a-half, making them the undefeated polo champions of California. Some 5,000 spectators watched the game... probably $20,000 changed hands."

Frank also built a beagle breeding facility to provide dogs for Sunday tally-ho events, and would be credited with bringing German Shepherds, animals who Carolan considered ferocious and closely related to wild tigers and bears, to California. Several memoirs and interviews noted that Frank and Harriett were their happiest at Crossways.

The next happenstance was an event of someone's misfortune turning into another's fortune. Pullman faced a strike that year. The strike cost his company $85 million, a sum it never recovered from. Shortly thereafter, he suffered a sudden and deadly heart attack. His children — Harriett, her sister and two male twins who became infamous for their drinking, womanizing and frivolous treatment of money — divided the huge estate with their mother.

An interesting $1.2 million bequeath from George's will was given to the Pullman Free School of Manual Training, and the presidency of the school was eventually turned over to successful attorney Robert Todd Lincoln, son of President Lincoln. This endowment was an odd connection, considering that it was Lincoln's death that had brought Pullman his fortune.

It wasn't long before the newly arrived and financially unburdened Mr. and Mrs. Francis Carolan were the toast of the Bay Area. Far from being treated as outside riffraff, as they may have been seen by the East Coast's standard of new money vs. old money, the two flourished in high society.

Harriett and Frank were known far and wide for their lavish social events at the Crossways, including one particular floral costume ball/garden cotillion in celebration of the completion of the Carolan's $200,000 clock-towered stables in 1900. It's reported that guests were transported from San Francisco to the Crossways in a string of Pullman coaches covered with flowers and potted plants. Costumed guests were met at the station by Frank's collection of elaborate horse-drawn carriages. The Crossways stable was lit with 5,000 paper lanterns and 3,000 electric lamps, and spotlights illuminated the manicured gardens.

Women were dressed in their finest gowns adorned with fresh-cut flowers. Harriett wore a pink gown with roses and a wreath woven with pink baby roses. Mining heiress Ella Baldwin wore a gown draped in pink carnations (the hybrid *Virginia Hobart* that had been named after her). Many of the men wore stylish pink hunting coats, no doubt made for the event, and white breeches. The champagne flowed and fine foods were served promptly as live music played and couples danced under the moon. This was the Peninsula at its best. The party lasted for six hours and was said to have cost $1,000 an hour. Harriett, known for her generosity, even had a second party for the hired help and their friends a week afterward. These parties were the dry run for what Harriett had planned for 1915.

The Panama-Pacific Exposition was only a few years away. The buzz was electric about the "competition" for the grandest estate to comfort and entertain the movers and shakers who would be traveling to the exposition. Ethel Crocker had already drawn up plans to add a ballroom to New Place (her grand estate on the Peninsula). Harriett, not one to be outdone, requested in 1912 that several French architects submit plans for her château on the hill. Again, Willis Polk began executing Harriett's plans. Construction on the Carolands estate began, but not without its problems, water being a main concern. The property on Black Mountain was overgrown and seemingly untamable. Without roads, construction was slow at best. The land, purchased from the Black Mountain Land and Water Company, had the best view on the Peninsula, but was difficult to reach. All of the concrete for the estate was mixed by hand with water brought in by slow-moving vehicles. Eventually, they traded property (for Eucalyptus Drive) with their neighbors, the Crockers, to build their throughway.

Although the Crossways Farm had survived the earthquake with relatively little damage, the couple's relationship was on very shaky ground. They spent most of their time away from each other: Frank was busy with politics while Harriett was often abroad in Europe or visiting her mother in New York. The years went on and the Carolands was nearing completion, though its blueprints grew with every grand palace that influenced Harriett during her trips through France.

In 1912 Pullman stock was at an all-time low, but construction on the Carolands was in full swing and could hardly be stopped because of lack of funds. The Crossways property was sold and Harriett and Frank moved into the Claremont Hotel in Berkeley until their new estate was completed.

A Greek temple of love was erected as well as fountains and a bandstand on the property's west slope. Although barns and bunkhouses, to house cows and ranch hands, were built, no garage to park Harriett's Pierce-Arrow and 16-cylinder Cadillac or Frank's 12-cylinder Cadillac was ever constructed.

The estate, located on a 500-acre lot, was finally completed near the end of 1914 and Harriett immediately began throwing extravagant parties that spared no expense. Soon the Vanderbilts, Roosevelts, Harrimans

The Carolands is currently under restoration.
Photo by Sally Richards

and Rockefellers had paid visits to the estate. Harriett had built an estate fit for nobility, and when she invited the world's movers and shakers, they came.

Although the house was immense, spacious enough to entertain 3,000 people with elbow room, it was not large enough to separate the couple. Harriett, becoming more withdrawn from her husband, was finding only momentary peace in her philanthropy work. One of her indulgences was her little dogs who became like her children to her. Sadly enough, or maybe for the better all things considered, she and Frank had never conceived a child of their own. On a trip to Paris she lost her favorite furry companion. Devastated, she returned home without one of the few things that brought happiness into her life.

The Parisian press picked up the fact that Harriett had lost her adored pet and was willing to pay a big reward for the return of her dog. Within weeks she received a telegram from a gentleman in Paris who had found her dog. She gave him $1,000 and expenses to personally accompany the dog back home via the fastest boat available. The *Chronicle* poked fun at Harriett's generous reward and ran a piece that claimed, "Mrs. Carolan is delighted over the return of her pet, especially so since its bluish coat exactly matches one of Mrs. Carolan's sets of furs."

Although the press was known to chide the Peninsula's wealthy, it was just as quick to praise them for their philanthropy. Harriett's generosity was known throughout the county. It was Harriett who originally held a campaign to build St. Matthew's Episcopal Church and donated the altar in memory of a friend.

Harriett was in Europe buying furniture for her new home when World War I broke out. She immediately put her fund-raising instints to work and helped form the American Ambulance Corps in France. Upon her return, she was named San Francisco's representative for the LaFayette Fund. In the meantime, Frank kept himself busy on the home front by fighting with local politicians and considered forming his own city to escape the cost of constructing a 3,000-foot sidewalk the city of Burlingame wanted him to put around his property.

The long-awaited Panama-Pacific International Exposition finally opened and both Harriett and Frank took part in organizing and promoting the event. In connection with the LaFayette Fund, Harriett gave a reading of Dana Burnet's poem "Albert of Belgium," with the addition of her own edits. Her reading made the *Examiner's* front page and caused months of agitation among San Francisco writers.

After America entered the war, life at the manse became difficult for the couple. The grand estate was far too small for the couple who preferred to be in separate

After the Panama-Pacific Exposition ended, the city watched as it was torn down, or auctioned off and relocated.
California History Center

corners of the earth, and Harriett's visits to Europe had ended for the time being. After the railways were seized and operated by the government, the couple's assets hit rock bottom. The home was stripped to a minimal number of servants but household expenses were still astronomical. High winds sent air back inside the French-cut chimneys, filling the house with a layer of soot. The French windows rattled constantly inside their panes and toilets and pumps made constant grumbling. Even on the calmest days, being inside the home was often compared to being aboard a ship at sea.

While Harriett busied herself by moving most of the furniture back East and checking into a New York hotel, Frank moved to the Fairmont Hotel in San Francisco where he died after a long illness. Harriett came back to San Francisco to fight a nasty court battle over Frank's will. It seems that many of the couple's documents were destroyed during the earthquake, and when Frank had the stock reissued, he had done so only in his name. Frank left all of the couple's assets to his friends and family. There was also the matter of a $125,000 vested life insurance policy taken out in Harriett's name. The court testimony made the front pages of Bay Area newspapers as she described that Frank had been a "$250 a week clerk" when she met him. By that time, there was only a little over $1 million in assets and the house to even argue over. The Pullman fortune was nearly exhausted.

Harriett did sell the house and after a number of successive owners, none of them seemingly happy in the home, the Carolands fell upon a stroke of luck. Once again, as happenstance would have it, there was another woman who loved French architecture as much as Harriett. The Countess Lillian Virginia Remillard Dandini acquired the estate for $80,000. This woman was more than a homeowner; she was also part of Carolands' history.

Today, at least before the renovation in completed, on the west side of the building, red bricks show through a cracked cement facade. These bricks were made by the factory that Remillard's father owned. The bricks, purchased by Willis Polk to build the Carolands, helped seed her father's fortune. Countess Dandini had a soft spot for the mansion.

"I never dreamed of having the house," the Countess once said in an interview. "But when I heard it was being torn down — it was too large and nobody wanted it — I bought it."

As with Harriett's father, Remillard's father, Pierre Remillard, had come to California to mine for gold. His plan fell through and he went to work for a brick company where he moved up the ranks and soon owned the business. The Remillard Brickyard in Oakland became a prosperous business. Bricks from the yard built St. Mary's College and the Phelan, Flood and Shreve estates.

The Countess, who also had extremely bad luck in love, might have followed her career as an opera singer, as she had just been invited to join the Metropolitan when

Tube courtesy of Vacuumtube.com
Photo by Sally Richards

her father died in 1904. Instead, she and her mother took over the company and became a force to be reckoned with. Happenstance struck again in 1906 when the earthquake made the Remillards instant millionaires. The Remillard's fortune grew as they purchased gravel pits and made wise investments. The company helped rebuild San Francisco and supplied the bricks for many famous landmarks in the area including the aggregate for the San Mateo Bridge.

The Carolands story does go on, but it is one of great travesty including a brutal murder and an heir holdup in the manse. As with many of the eccentric characters of the colorful era, the mansion ceased to be a place of importance and settled into the background while the next generation began making its own history.

Invention Meets the Entrepreneur

The universities in the Bay Area, Stanford, Santa Clara, UC Berkeley and San Jose State, continued to feed young, educated dreamers into the local work force. With each new graduating class, the Valley's sophistication and knowledge grew and many alumni went on to discover new technologies and form companies that continue to generate jobs and income for the Bay Area.

Back in the days of old, "wireless" communication had quite a different reference than it does today. Shortly after 1900, a wireless communications device based on radio technology was invented in Europe and soon several American wireless companies opened up shop in San Francisco to establish ship-to-shore communications. Entrepreneurs went to work to develop radio technologies and were only encumbered as to how to negotiate around Guglielmo Marconi's (the inventor of a successful wireless communications device) patents.

In 1906 Cyril Elwell, a graduate of Stanford's electrical engineering program, found a way around Marconi by purchasing American rights to patents held by Vlademar Poulson, a Danish inventor who also held patents in wireless technology. Stanford President David Starr Jordan encouraged several staff members to invest in Elwell's company, Federal Telegraph, and they did. One of the company's researchers was Lee de Forest, who holds the title "Father of Radio." It was de Forest who used "Edison's Effect" in his development of the three-element vacuum tube to which he gave the catchy name Audion.

Later, the company's manufacturing plant would be located in Palo Alto and many of Stanford's engineering graduates would begin their careers with Federal. Before

Lee de Forest
California History Center

Federal would leave the Bay Area in 1932 for its new East Coast home, it generously gifted UC Berkeley an 80-ton, 1,000-kilowatt arc generator. The behemoth machine was used by Ernest Lawrence to build his cyclotron device.

In 1910 an arrival from Indiana would change the course of history. A boy by the name of Frederick Terman was brought to live at Stanford by his father, Lewis Terman, a psychologist who had joined the Stanford staff. Fred Terman was steeped in the academic atmosphere of Stanford from the time he was 10 years old. He grew up around some of the greatest minds in the country, and some of it must have worn off on him because by the time he reached 16 he had helped to build an amateur radio transmitter with his boyhood friend Herbert Hoover, Jr. He was a young dreamer who learned early on from his many mentors on campus that cooperation was key in building great things.

Naturally Terman chose to attend Stanford. He graduated at the age of 20 with a degree in chemical engineering. As was the tradition of many Stanford graduates, he went to work for a brief time for Federal Telegraph in the research department. Terman had a thirst for knowledge and soon enrolled in the electrical engineering program at Stanford where his father's close friend Harris Ryan mentored him. After graduation in 1922, he decided to find out what the great minds on the East Coast were working on, and enrolled at MIT where, two years later, he received his doctorate in electrical engineering. Terman accepted a staff position at MIT but first came home to visit his family and friends before starting his teaching career. While there Terman was diagnosed with tuberculosis. While recuperating at Stanford, Ryan offered him a part-time teaching position.

Soon thereafter, he began his teaching career at Stanford that would span decades and play a critical role in the university's future success. Terman began recruiting the brightest students for research projects in vacuum tube and radio wave technologies. The Roaring 20s were a great time for the growth of the electronics industry in the Valley and for Terman personally, who felt the key to a successful learning institution was the marriage of commercial and academic research. He began building relationships with people from both camps.

The fun-filled times of the 20s ended suddenly in 1927 when the Great Depression hit America. Ryan

retired in 1932 and Terman became a pivotal part of the progress of the engineering department. Also that year, Federal Telegraph, one of the Valley's largest employers, packed up stakes and moved to the East Coast.

The Great Depression of the late 20s and 30s devastated the entire United States, but government was doing all it could to stabilize the economy. People learned to make do with what they had. The Bay Area's wealthy, who had become spoiled during the play days of the Roaring 20s, were forced to let their servants go, sell property and cut back on their trademark eccentricities and expensive habits. The Bay Area took on a somber lifestyle, bottling up much of its joy. Soup lines in Bay Area cities stretched for blocks and no one was spared the ugliness the Depression took on.

The Depression brought about many changes in the banking industry and social programs that would affect Valley residents and the way they managed their money and lives. The movement toward a more stable and safer America began in March 1933 when all banks were closed and only those receiving approval from federal investigators were allowed to reopen. This trend also saw the beginnings of the FDIC (Federal Deposit Insurance Corporation), the FSLIC (Federal Savings and Loan Insurance Corporation), the SEC (Securities Exchange Commission), the NLRB (National Labor Relations Board), FLSA (Fair Labor Standards Act) and Social Security.

The FERA (Federal Emergency Relief Association) began doling out direct relief in the form of cash to aid states and localities for distribution to needy. Ultimately, FERA distributed about $3 billion in relief to 8 million families, one-sixth of the United States' population. The CWA (Civil Works Administration) kept idle men working when it hired workers to construct its planned 225,000 miles of U.S. roads, 30,000 schools and 3,700 playing fields and athletic grounds. The PWA (Public Works Administration) facilitated loans to private industry to build dams, ports, bridges, sewage and power plants, airports, hospitals and government buildings.

One of the Bay Area companies that flourished at this time was the Permanente Cement Corporation, a company owned by Henry J. Kaiser. Originally Kaiser was just a man making a living and pushing his small company to grow in any way he could. It wasn't until 1927, when he was subcontracted to build 200 miles of paved road in Cuba, that he finally got his big break. Kaiser built the road a year ahead of schedule and made such a large profit that he was able to go directly into the dam building business, eventually constructing the Shasta, Hoover Bonneville and Grand Coulee dams. 1939 saw a tremendous growth in the company, and the construction during World War II helped build the Kaiser empire.

The medical insurance industry was a major hurdle for employers who wanted their workers to stay healthy enough to complete big, dangerous projects such as dam building. Employees were also taking a huge risk on these dangerous jobs where they could be seriously injured and have no hospital care nearby, and no way to pay for it if there was. During the construction of the Los Angeles Aqueduct an ambitious doctor established a temporary hospital in the Mojave Desert to serve thousands of men. The project had just ended and Dr. Sidney Garfield was about to throw in the towel because of the difficulties he faced collecting on insurance claims. His 12-bed hospital had not been very successful. Harold Hatch, an engineer-turned-insurance agent, approached Garfield about a plan that might make both of them very rich. Hatch suggested that insurance companies pay a specific dollar amount per day, per worker — in advance for coverage. The prepayment plan was instituted. For the sum of five cents per day, workers received full coverage for on- and off-job injuries and illnesses. Thousands enrolled and the news of Garfield and Hatch's plan spread far and wide, right to the ears of Henry J. Kaiser.

Kaiser, the czar of cement, made a proposition. Could Garfield and Hatch insure and care for the 6,500 workers and their families at the largest construction site in history — the Grand Coulee Dam project? Up to the challenge of the massive project, Garfield put his solo-practice plans on hold, turned his existing run-down hospital into a state-of-the-art treatment facility and recruited a team of doctors to work in the hospital. As the dam neared completion it seemed that Garfield would go back into private practice, but World War II intervened.

The war's needs brought thousands of ship workers to the Kaiser Shipyards in Richmond to meet the contracts for big Liberty Ships and aircraft carriers. Kaiser again had the problem of providing health care for this teeming mass of 30,000 people. Although Garfield was slotted to enter active duty with his Army reserve unit in a few weeks, Kaiser asked a favor of President Franklin Roosevelt to release Garfield from his military obligation so he could organize and run a prepaid group practice for the workers at the Richmond shipyards. Roosevelt granted Garfield immunity from military duty and his prepaid plan eventually became a permanent part of Kaiser's empire.

Other programs implemented during the Depression and into World War II included the CCC (Civilian Conservation Corps), an organization responsible for many of California's parks and upkeep of its open space, founded to provide men between the ages of 18 and 25 with jobs under the direction of the U.S. Army. The WPA (Works Progress Administration) was set into motion to provide jobs to men who would repair public facilities. The WPA plan also included the Federal Arts Project that provided pay for artists who could provide drama, writing, music, murals, sculptures and other forms of art to the public.

LOOKING FOR ANSWERS FROM ABOVE

During the Depression, the Valley's cities were looking for opportunities to bring more work and disposable income into the Valley. Oddly enough, the answer came from above, in the form of two rigid dirigibles belonging to the U.S. Navy. The Navy relied upon the strategy of using lighter-than-air ships to build its defenses. The idea had originally come from the Germans who had utilized them successfully during World War I. Where would the Navy house the monolithic air vehicles? It was going to take wide-open space and a cooperative community. Although at first the Navy was in favor of building the dirigible hangars at Camp Kearney in San Diego, James Rolph, Mayor of

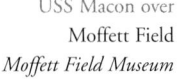
Practicing war maneuvers over Moffett's salt marshes
Moffett Field Museum

San Francisco, made a plea to the citizens of the Bay Area to help find the base a home. It was young real estate agent Laura Whipple of the East Bay who envisioned the base on the old Ynigo Ranch between Mountain View and Sunnyvale. Later, she was asked what she thought when she first saw the 1,700-acre site covered with row crops. She answered, "I saw an air base."

Whipple began a campaign involving the San Jose Chamber of Commerce and local press to persuade the U.S. Navy to bring the base to the Peninsula. Cities far and wide joined in the rally, as they would do later for the construction of the Golden Gate Bridge in 1937. In an attempt to seal the deal, the Bay Area's four counties raised the sum of $476,066 to purchase the land and offered to donate it to the Navy if it would commit to building the base on the property. Two months later President Herbert Hoover, a Stanford graduate, signed the paperwork that would give the Navy the authority to accept the property for the sum of $1. Hoover appropriated $5 million for the construction of the base. The base's construction provided hundreds of construction jobs and a need for millions of dollars in building materials from local manufacturers, not to mention the income the sailors would continue to feed into the surrounding communities.

On the foggy morning of May 13, 1932, an estimated 100,000 people lined up along the Bayshore Freeway, under construction at the time, to witness the unbelievable event of the silver USS ZRS-4 Akron passing overhead. It would be the last time the vehicle would come to the Valley, as it was stationed in New Jersey until the air station was completed the following year. Later in 1932, the

USS Macon over Moffett Field
Moffett Field Museum

Akron crashed off of the Atlantic and 73 of the 76 crew members were killed, including Adm. William Moffett whom the station would later be named after. In the end, the 211-foot Hangar One landmark could be seen from just about anywhere in the Bay Area, a testament to what the Valley's cities could do if they worked together. 1933 saw the arrival of the Akron's younger 785-foot-long sister, the ZRS-5 Macon, the airship that would call Moffett Field home.

The dark shadows cast over the orchards by the Macon as it moved slowly against the Valley's blue sky as it practiced its war maneuvers was only one of the many preludes of war. The future of dirigibles was unknown because the government was constantly testing other means of defense, but all involved with Moffett knew they needed to prove their worth to keep the project funded.

One such event occurred in July of 1934 when President Roosevelt was on a journey aboard the cruiser *Houston* sailing through the Panama Canal en route to Honolulu. Using only what news of the president's departure they could find in the papers, the Macon's commanding officer raced 3,500 miles toward the Pacific Ocean where he determined the *Houston* would be. Crewmen aboard the *Houston* were caught totally off guard when the Macon released two scouting airplanes to circle the ship. The Macon descended with great drama from the clouds where it was hidden to drop another plane from its hollow belly. The small reconnaissance plane released bundled stacks of the previous day's Bay Area newspaper onto the *Houston*'s deck. From later accounts, it was clear that the fleet's admirals were not amused. The president, however, found the dramatic antic quite entertaining.

Blimp on carrier
Moffett Field Museum

During the war Moffett was quite active with its full fleet of blimps that patrolled the Pacific in search of approaching enemies. However, the technology of the rigid dirigibles was not meant to be. Moffett Field, a name recognized by many Valleyites despite numerous name changes, faded in and out of the Valley's history many times. One of Moffett's most famous young cadets was Hollywood icon James "Jimmy" Stewart. The handsome star stayed less than a year and eventually rose in rank to major general in the Air Force Reserves. Over the years, thousands of military men and women would serve their time and upon discharge join the Valley's civilian work force.

Moffett Field drew the attention of many companies that wanted to be included in the activity the base would provide. Soon, the open land around the base was filling with companies. The Peninsula was a hot spot for new companies because of its location to Stanford, and in 1932 it was fertile ground for the vacuum tube industry. Companies such as Charles Litton's Litton Electronics and Eitel-McCullough were going strong. The vacuum tube was the new king of the Valley.

Meanwhile, Bay Area scientists and engineers were making great strides in technologies that would aid war efforts. Other products developed for the war would also later become commercial products that would change the world.

A MADMAN UNITES HIS ENEMIES

1933 marked the year Adolph Hitler came to power as the dictator of Germany, an appointment that set into motion a series of events that would bring the entire world into war. The American government, instead of relying on the number of men in its armies, urged and underwrote scientists creating cutting-edge technologies that would give the United States an edge over its enemies as it readied itself to enter the war.

By 1935 Hitler denounced the provisions of the Treaty of Versailles and reinstituted compulsory military service. Soon thereafter, Italian Dictator Benito Mussolini began his invasion of Ethiopia. In an effort to keep peace, the British and French governments tried to intervene with a compromise agreement, but by 1936 Italy had completely taken over Ethiopia. World powers took note of Germany's massive rearmament and began stocking their own weaponry. Finding strength in numbers, France joined forces with the USSR. Hitler retaliated by remilitarizing the Rhineland. Britain and France could have taken

Recon blimps stored in the hangar
Moffett Field Museum

Czechoslovakia and an important port in Lithuania that pushed Poland into a corner. The shoving match began. On September 1, 1939, Britain and France swore their allegiances to Poland and two days later declared war against Germany.

Meanwhile, the threads tying America's immigrants to their relatives living in Europe began to unravel. Small foreign wars and border disputes took on larger meaning as America, a land that once seemed isolated from the threat of foreign attack, slowly realized its democratic freedom could also be at stake. Stories of the war's atrocities and a madman named Hitler began to trickle in until the war was carried on the front pages of America's papers. Unable to ignore its obligations as a major power, America's government continued to fill its war coffers in anticipation of joining the war as it scoured universities and private industry to find the technology needed to break Germany's grip.

Germany in that moment of time, but instead opted to try to peacefully resolve their differences with Germany.

Heady with his success, Hitler took over Austria in 1938. Momentarily satiated with their progress, Hitler and Mussolini agreed to listen to French and British officials

BUILDING THE MEANS TO COUNTERATTACK

In 1938 two brothers named Russell and Sigurd Varian changed the course of World War II. Russell, a determined and intelligent visionary who received his degree in physics from Stanford, had a keen sense for developing scientific concepts into product.

Earlier in his career Russell had cut his research teeth while working with inventor Philo Farnsworth to help develop the technology that would later become television. Farnsworth's method of electromagnetically focusing and deflecting electron beams was a winner. Farnsworth, a college dropout, went through many patent fights with RCA and would eventually lose everything.

In America, life went on and business was good.
Redwood City Public Library Local History Room

at a peace conference in Munich. Hitler agreed to a nonaggressive pact with the USSR. In complete defiance of the agreement he secured the remaining free areas of

On September 7, 1927, Farnsworth wrote in his journal: "The received line picture was evident this time." Thus, the television had its birth. RCA's David Sarnoff hounded Farnsworth to give up his patents for $100,000, but Philo knew their true worth. All of Farnsworth's television patents were running out during the war as commercial efforts were put into making weapons and developing technology to ensure America's success. Sarnoff knew if he waited out the war, the technology would be his. Sarnoff aligned himself with Russian immigrant Vladimir Zworykin, an inventor who was working on quite a different TV principle, until he made a friendly visit to Farnsworth's laboratory.

As a result of the visit, Farnsworth returned to his farm in Maine a broken man. The Father of Television drank heavily and became addicted to painkillers. Sarnoff triumphed and Farnsworth became a footnote. Johnny Carson's eulogy of Farnsworth upon his death in 1971 was, "If it weren't for Philo T. Farnsworth... we'd still be eating frozen radio dinners."

The next luck of the draw was dealt to Sigurd and Russell Varian. Sigurd, a tall cowboy-like pilot with the heart of an adventurer, a booming voice and a wonderful sense of humor, convinced Russell to return home, 250 miles south of the Valley, to open a workshop where they could invent a radar device. As war activities in Europe increased, Sigurd, a pilot with Pan American who flew the American border area via the canal zone, was adamant that he and his brother could develop a state-of-the-art radar system that would change the course of the war.

Sigurd and Russell put their heads together and developed the beginnings of what would be a major advancement in radar. Realizing they had to move closer to Stanford to work with William Hansen, Russell's old roommate who was working on complimentary technology and had joined the Stanford chemistry staff, the brothers set up shop in San Carlos and later moved to Stanford. The Varian building was the cornerstone for what was to become the Stanford Industrial Park.

Much of Stanford's success with its industry ties is due to a decision made long before even Terman came to the university. When the Stanfords endowed the university, they did so with the clause that the land, all 8,800 acres, would never be sold. Although parts of the property have been taken via imminent domain, it has remained relatively whole. One of the ways Stanford has been able to leverage the property to increase its revenues is by leasing the property to companies that it wishes to be strategically aligned with.

The university funded some of the Varian brothers' development costs and gave them space in its lab, and by doing so Stanford also shared in Varian's proceeds. The company developed the klystron and by the late 40s, just as the Luftwaffe had begun deadly nighttime bombing raids, the Royal Air Force completed equipping its night fighters with the klystron radar receivers that would ensure success in the Battle of Britain.

Klystron Tubes made by Varian
Palo Alto Times Collection
History San José

"In fact, it was his urging rather that anything else that started the project off," said Russell Varian about his brother in a 1958 taped interview that is housed in Stanford's archives.

The interview, conducted at Varian's 10th anniversary party, was the event also marking the 20th anniversary of the invention of the klystron tube. "He had done a great deal of blind flying and night flying. I think that when the war started the Germans had, if anything, a little bit better radar than anyone else had. It was long-range radar — this big cumbersome thing that would be good for early warning. But the whole philosophy of the Nazi attack was that the war was to be a short war and any further development was useless. So, they pushed practically all of their strength into production of equipment.

"When it became obvious that there was not to be a short war, then they started to work in military development. As far as I know, they didn't do very much in radar. They were so far behind at that time that they didn't do much other than barely try and counter some of the measures being used.

"It proved to be rather easy to persuade a lot of people who were well trained and had a lot of valuable experience to come back to the West Coast," said Russell Varian of his successful recruitment of East Coast talent for his company. "That's one of the factors [in Varian's success].

Photo by Sally Richards

Another [factor] was that the firms developing on the West Coast were mostly developed from the direct action of engineers. There had been a great deal of frustration on the part of engineers and scientists with the management of some of the larger companies that didn't understand what was required to do effective research. They [the large East Coast companies] were committed to doing research and would appropriate the money, but usually they would administer it in such a way that it was difficult to make progress. I think possibly one of the great reasons for the advance [in technology] here [in Silicon Valley] is that it was a fresh start."

Also attending the event was Air Vice Marshall Lang of the Royal British Air Force who commended the Varian brothers on their invention:

"As the first director of radar in the London, after the beginning of the war, I was intimately concerned with the operational value and application of radar. Our problem of intercepting German aircrafts was difficult, at that time in early 1939 [we] only had a few radar stations and were using long wavelengths. We were badly in need of the greater improvements that would give us the greater precision necessary to make the air power that much more effective. The advent of the klystron, the invention of the Varian brothers, filled this gap. We needed badly to get into the microwave field and the klystron made this possible. The advances that became possible were revolutionary from the standpoint of increasing the fighting power of the Royal British Airforce, the navy, the anti-aircraft armaments and the coast artillery. Tremendous advances were made through 1941 by applying the klystron to new techniques, and the tremendous efforts of this country to get into high-speed production of this valuable edition of technology, are what eventually made sure air power was overwhelming and it resulted in combined victory."

The klystron also turned out to be much more than a critical wartime tool; it would also play an important part in developing an industry known as microwave. The klystron would go on to make commercial air navigation safer and develop into high-energy particle accelerators that would change the industries of medicine, nuclear physics and worldwide communications via satellites.

One of the most unsung good deeds of Russell Varian was his dream of purchasing the once lumber-stripped mountains high above Silicon Valley. Russell, an athlete and hiker, often wandered the mountains for inspiration and wished to protect the mountains from ever being stripped as they had been for logging in the 1800s. Russell pushed to rally the state to purchase the property but they had no interest. Russell took matters into his own hands and moved to purchase the property himself. In 1959, after he received an option to buy his cherished 27 acres, now the heart of Castle Rock, he died before the purchase could be completed. Later that year, memorial funds sent by his friends and associates secured his 27 acres, and in 1968 the State Park and Recreation Commission designated Castle Rock as a state park. The area, now a haven for the Valley's rock climbers, includes 3,600 acres of the wilderness that Russell Varian felt was Silicon Valley's responsibility and honor to protect.

THE BOMB'S SILENT CASUALTY

The ocean's warm waves washed back and forth against carriers and submarines of the U.S. Navy's Pacific Fleet protecting the island of Oahu on the morning of December 7, 1941. It was early dawn, the sun's rays barely reaching over the horizon, when the naval aviation forces of the Empire of Japan caught the island off guard and unleashed bombers on its sleeping Pearl Harbor target. Japan's goal of attacking was to totally annihilate the U.S. Navy's Pacific Fleet so it could capture the Philippines and Indo-China. By doing so, it would secure the raw materials needed to ensure its success as a global power. Japan's plan was to extend its empire to include Australia, India and New Zealand. Japan expected Germany to take Great Britain and Soviet Russia, and its own control of

San Jose in the late 30s
History San José

the Pacific. Had events shaken out the way Japan saw them, Japan would have controlled East, Southeast and South Asia; Germany and Italy would be in control of Great Britain, Europe, Western and Central Asia, the Middle East and Africa; and the United States would rule North and South America. Instead, the secret air attack brought America — and its atom-splitting technologies — into the war.

Just after the time the klystron was in development, news had reached the United States that German scientists had split the atom. In 1941, realizing that individual scientists were not working fast enough on splitting the atom, President Roosevelt established the Manhattan Project. Also in 1942, a scientist named Robert Oppenheimer was appointed to direct the efforts in building a super bomb before the Germans did. Much of the research had been done at Columbia University, the University of Chicago, and in Oak Ridge, Tennessee, but Oppenheimer looked toward the desert in anticipation of testing his creation, and set up the project in New Mexico. He, with additional persuasion from the president, brought together more than 3,000 workers and physics geniuses at Los Alamos to create the atomic bomb.

An event that occurred on July 16, 1945, finally put into motion the end of the war. On that fateful day, Oppenheimer witnessed the first explosion of the atomic bomb as it mushroomed into New Mexico's desert sky. "We knew the world would not be the same," he later commented. Within a month, two atomic bombs were dropped on major Japanese cities, and Japan surrendered on August 10, 1945.

After the war, Oppenheimer chaired the U.S. Atomic Commission and at Truman's request, and much to his dismay, he began work on the hydrogen bomb. In a strange twist of fate, Oppenheimer's moral reluctance to work on a product of even greater destruction turned the president against him. In 1953, while all those around him were being accused of communist activities during the anticommunist trials, he was taken down and stripped of his clearance. The truth be told, he did have friends who were communists, mostly intellectuals involved in the anti-fascist movement of the era. The same witch-hunting committee that held trials destroying artists, producers and writers such as Dashiell Hammett found Oppenheimer guilty of having communist sympathies. Oppenheimer turned his back on the government and went back to teaching, this time at Princeton as the director of the Institute of Advanced Study, a chair once held by Albert Einstein.

He had many regrets later in life about the bomb he created and wrote many papers on the subjects of morality and ethics. It was clear Oppenheimer had much to say about the events set into motion with his atomic bomb. Ironically, he died of throat cancer in 1967.

War Brings Prosperity to the Valley

If war can ever be considered for the unity and technological advancements it brings about, then despite the bloodshed, World War II must be credited with the positive changes it brought to the Valley. The government looked to the technology being developed in California to create weapons that would ensure world domination. One company directly linked with the Valley's past and its future is FMC, Food Machinery Corporation, a business founded in 1883 by a retired farmer who designed and sold farm equipment. The company grew and acquired another growing company and was soon a serious contender in the machine industry. During World War II the company diversified its product line and its engineers developed the Water Buffalo, an amphibious tank-like vehicle.

The Valley is filled with stories of companies

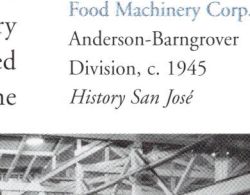

Food Machinery Corp. Anderson-Barngrover Division, c. 1945
History San José

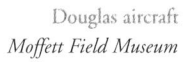

The first day Ampex opened its doors for business
Redwood City Public Library Local History Room

that were founded to meet the war's needs or strategically reinvented themselves to develop the new market. The Depression was over and during the war there was work for everyone, including women who needed to fill the jobs men had abandoned to fight on the front lines. When the war ended, men, of course, wanted their jobs back, but women's roles in industry would be forever changed. Enrollment of women receiving scientific and engineering degrees from top universities began to steadily increase. The next four decades would prove women equal in the workplace and propel the movement of equal work for equal pay. With time, changes in social attitude and much-needed confrontation, women would eventually make incredible strides to ensure their places in corner offices and boardrooms.

The government, not willing to forget the lessons learned during the Depression, pushed forward its beliefs about free trade. Government and industry rallied to eliminate trade barriers and create markets for American agricultural and industrial products. After the war, the Soviet Union had its own ideas about free trade. The historical Russian tradition of a centralized, autocratic government went head to head with America's beliefs about democracy. The Marxist-Leninist ideologies that had taken a back seat during the war were still the basis of the Soviet government's policy. The Soviet Union had lost more than 20 million people during the war and

planned to build an even higher wall to prepare for the next conflict. The Soviet government became suspicious of another attack, one it felt would surely be launched from its superpower equal in the West. This was the beginning of the Cold War, a time when the U.S. government would turn to science and technology to protect its borders and promote democracy abroad.

If war infused prosperity into the Valley's economy, the threat of the next war was even better for the Bay Area and California in general. By this time, Terman had been named dean of engineering and would eventually go on to become provost, and worked even harder to promote the partnerships between Stanford and industry.

THE COMING AGE

The Cold War caused a whole new defense industry to flourish, and it was common to see engineers on their way to work and women on their way to administrative and manufacturing jobs in the morning. The Valley had a new mission — to protect America's borders.

Before the war's end, a plan Terman set into motion would begin to gel. He brought together David Packard and Bill Hewlett, both graduates of the engineering department.

Born September 7, 1912, in Pueblo, Colorado, Packard came west to the Valley to

Douglas aircraft
Moffett Field Museum

meet others who were as curious as he was about the new technologies that would set the bar for the rest of the world. He had a gift of remaining quiet and listening to what anyone said before commenting. Packard and Hewlett set up shop in the famous garage of Bill Hewlett's house in Palo Alto, today a national monument where tour buses stop to show visitors where Hewlett-Packard (HP) started. The two entrepreneurs worked hard to develop a product that would change the way the world was entertained.

When Charles Litton had patented multi-grid vacuum tubes in which oscillations occurred, because of Stanford's involvement in the company the patents were granted to the university. Then when the university received $1,000 from Sperry Company for use of the patents, Litton insisted that the money go into the engineering department for electronic tubes. Agreeing wholeheartedly, Terman signed over a check directly to Packard, who was signed up at Stanford for graduate work, for the amount of $1,000 for product development. With Terman's guidance, they created the commercial audio-oscillator, a product that generated signals of many different inaudible frequencies to analyze distortion. The dynamic duo sold eight to Walt Disney, who was building a kingdom that would eventually seep into nearly every facet of retail and communications presently known. Walt Disney's mouse helped build the HP empire when its oscillators were used in the production of *Fantasia*.

In a product twist that illustrates the times, after the Disney contract, the company adapted oscillators for the war industry. The Korean War created new needs in technology, and HP was making sure its products diversified to meet those needs. The products they created included radar and counter-radar equipment and microwave signal generators. Its product line grew, and soon the company was out of the garage and into a building that by the turn of the millennium would employ more than 100,000 employees worldwide. Naturally, HP became one of the growing number of companies in Stanford's Industrial Park.

Not only did Hewlett and Packard develop an industry, they also brought a humanity to the workplace known as "The HP Way." This new management style encouraged creativity and their employees participated in the most generous benefits and stock plan at the time (which is still one of the best). Many companies try to emulate the management style developed by Hewlett and Packard, but few have managed to do so successfully. In 1996 the late David Packard stepped down as chairman emeritus at HP and, in true Silicon Valley entrepreneurial style, bid a fond farewell: "I think you get the most satisfaction in trying to do something useful. After you've done that, you ought to forget about it and do something else. You shouldn't gloat about anything you've done; you ought to keep going and try to find something better to do."

Continued Industry Evolution

Of course, not every technological advancement was discovered on the West Coast. The East Coast had its own set of geniuses who were hard at work making the world a different place.

In 1946 at the University of Pennsylvania the Sperry-Rand Company created a monolithic machine of calculation dubbed the ENIAC (Electronic Numerical Integrator and Computer). Looking back on this primitive great-grandfather of the modern computer, this monster of a machine that only had switches — and no monitor — it's difficult to believe that it could in any way change the course of history. But it did. The ENIAC was the first electronic computing machine, its technology based on a mind-boggling 18,000 vacuum tubes, 70,000 resistors and 5 million soldered joints. The computer was such a massive piece of machinery that it consumed 160 kilowatts of electrical power.

Developed by John Presper Eckert and John W. Mauchly, the machine switched to either the on or off position that translated into a yes or no answer. Today this language is known as binary code. It was a great first attempt, but what was really incredible about this machine was the number of engineers who saw it, liked it and knew they could do better. In essence, the ENIAC was the catalyst for a new way of thinking about computing.

Poor Management Leads to the Birth of Silicon Valley

In December of 1947 a breakthrough occurred when three engineers from AT&T Bell Telephone Laboratory, Walter Brattain, John Bardeen and William Shockley, demonstrated the principle of amplifying an electrical current using a solid semiconducting material. Their theory, based on selective control of the flow of electricity through germanium, designated some areas as current conductors and adjacent areas as insulators. This is where the word *semi*conductor is derived from. These inventors found a suitable alternative for the vacuum tube. Although tubes could be used for circuit switching, voice amplification, electromechanical and other important tasks involving the regulated conduction of electrical current, they were still very unreliable. The trio helped form the technology behind the transistor, the more reliable and much more easily distributed electrical transfer resistor. This little invention would change technology.

Ampex received several Oscars for its sound technology. *Redwood City Public Library Local History Room*

Before the transistor, Arthur C. Clarke's geosynchronous communication satellite was simply fiction, an inspiration for impressionable teen-agers who spent countless hours soldering their own electrical monstrosities together. Clarke's satellite, filled with millions of vacuum tubes, would have been the size of Willow Glenn and staffed with hundreds of employees whose only job would be to systematically replace the tubes as they blew out. Many of today's technologists of high tech look back and credit Clarke's work for their inspiration, including one young girl who grew up to be Donna Shirley, the engineer who would manage the Mars Exploration Project at the Jet Propulsion Laboratory (JPL).

In the Valley, just before and after the 50s, Lockheed, Ampex, Ford Philco, General Electric and Lockheed's Space and Missile Division worked full-speed ahead. Military buzz haircuts, television, cars the size of mid-size boats and rock and roll were working a solid foothold into California society. The future for the defense industry looked promising and investors were hot for any kind of technology that could shoot missiles and rockets, shoot down missiles and rockets or help put man on the moon.

Life in the Valley was getting faster. The construction of freeways, expressways, bridges and streetlights were keeping public works busy. Car salesmen standing outside of dealerships continued to feed automobiles into the Valley.

In 1952, three years before Shockley would strike out on his own to make history in the Valley, IBM, a company based back East, introduced the IBM 701, a large computer capable of executing 17,000 instructions per second. The 701, based on vacuum tube technology, was primarily used in government and research work. The tubes were quicker, smaller and more easily replaced than their previous models. IBM soon moved its product line into business applications such as billing, payroll and inventory control.

Not wanting to lose out on the technological advancements being made in the Valley, in 1953 IBM set up a research lab in San Jose. The scientists at the lab founded RAMAC (Random Access Memory Accounting System). RAMAC was, at the time, quite a step forward. This technology brought a new industry to the table: DASD (Direct Access Storage Device). IBM would become the king of storage and rule its domain with an iron fist. The company clearly excelled in sales and distribution. In a few years it would flood its mainframe machines into government and commercial markets worldwide.

In 1955, Shockley, a young man from of a long line of bluebloods traceable to the *Mayflower*, left Bell Laboratories to strike out on his own. He was born from parents who knew the value of an education — his father, an engineering graduate of MIT, was a mining engineer, and his mother was a mining surveyor and one of Stanford's first women graduates. He was no doubt a serious child who developed a respect for the elements at an early age. He received his Ph.D. in 1936 from MIT; his thesis addressed the energy band structure of sodium chloride.

When Shockley, Bardeen and Brattain developed the transistor, it was a piece of gold foil wrapped around a plastic knife and pressed against a block of silicon that had an electrical connection at its base. Compared to today's prototypes, the trio's device was archaic, but they had invented a switch to control electricity flow and the power to amplify it. This switch allowed work to be done at the speed of light. A whole new world opened up for engineers who would build upon this theory. The boys from Bell picked up numerous commendations for their creation, but none that allowed them to create more technology than the $38,633 they each received for winning the 1956 Nobel Prize for physics.

Shockley had other defining moments in his career, including his early work during World War II when he and a colleague developed a nuclear device independent of the Manhattan Project. Shockley's strategic analysis of the effects of aerial bombing perhaps even contributed to the decision to drop the atomic bomb on Japan.

With the war behind him, Shockley packed up and went west. He would set up a semiconductor lab at Beckman Instruments and then his own company, Shockley Transistor, Co., where he would hire brilliant engineers who would change the world. It seems now that Shockley had another title, Silicon Valley's first micromanager. He was home schooled most of his life, and some believe he was unable to build the socialization skills necessary to form equal relationships or grow a company of loyal employees. In other words, maybe he never learned to play well with others.

Apparently, he treated his brilliant team with disrespect and followed their every move, even asking them to take in-depth psychological testing before hiring them. Later, when he taught at Stanford, he started controversy as he left his engineering work to speak out on the topic of eugenics, specifically race and IQ. His theory was that blacks consistently scored 10 to 20 points lower than whites on IQ tests. His theory angered Stanford students enough to burn Shockley in effigy. For nearly a decade, his own colleagues from the National Academy of Sciences refused his requests for a hearing, telling him instead to publish his findings in scientific journals for peer review. Shockley is now a footnote in Silicon Valley's history, his most recognized claim to fame being the mismanagement of his hand-picked Shockley Transistor team, which caused their mass exodus. His poor management skills birthed the semiconductor industry that blossomed into one of the world's most important technological evolutions.

After Shockley Transistor, Co. was formed, life-changing events happened in short order. Eight of his talented engineers, hired to build the basis of the scientific theories that would create the semiconductor industry, walked out en masse. Shockley's management style, or lack thereof, had stifled the creativity of his team. The fed-up engineers included Gordon Moore, C. Sheldon Roberts, Eugene Kleiner, Victor Grinich, Julius Blank, Jean Hoerni and Jay Last, and would later be joined by Bob Noyce. It was 1957 when the crew Shockley would publicly call "the Traitorous Eight," formed another company.

"His management style, to say it kindly, was original," said Eugene Kleiner, one of the original eight. Vienna-born Kleiner would go on to a legendary career in venture capital. "He tried to micromanage some very creative people. I've worked with different kinds of managers before — some were good, some were not good. He was unusual about the way he handled things. He tried to punish people and it was very unusual. He almost treated us like children. It just didn't work."

When the eight walked out, they did so without funding. Feeling that the unknown was still better than working with Shockley, they began looking for a venture capitalist to fund their research. The youngest engineer in the group was 28, the oldest 34.

The early days at IBM
Intel

Eugene Kleiner
Photo by Sally Richards

"When we tried to get funding for starting a new company, we basically wrote a very unusual letter. It basically said 'Here we are. We don't like the Nobel Prize winner we are working with, but we like each other. Would you please give us a few million dollars so we can start a company?' It took some doing but we did get it," says Kleiner, smiling as he thinks back on the tremendous risk the group took. "Nobody on Wall Street knew much about semiconductors, so they really couldn't judge too much about the product. They just wanted to meet us."

Young Arthur Rock, who later went off on his own to became an icon in venture capital, found a good match for the young engineers. Fairchild Camera and Instruments wanted to branch into new technologies, so they funded the group. Fairchild Semiconductor was registered as an independent California corporation. In a move that showed Fairchild Camera and Instruments knew the importance of the future of semiconductors, it included a buy-back option attached with the capital. Nevertheless, thrilled with the opportunity to create freely, the eight jumped right into the fray of the burgeoning industry.

"Nowadays, you need a big checkbook so you can buy all the equipment," said Kleiner of birthing a new industry. "At that time the equipment was not available and we had to design and build it, but that was half the fun. Especially for myself because I'm not a semiconductor genius, but I knew how to build equipment; that was part of my responsibility. But we all worked on the problems together. It was not a case of trial and error, in other words there was a scientific approach to the whole thing. That doesn't mean we didn't make any mistakes. Decisions were made on a very scientific basis... it was relatively smooth going."

In 1958 Fairchild built its first device for IBM, a core memory driver that went into an airborne computer. The team, having used germanium on the first round of developing the product, found it couldn't withstand the heat. The team discovered that silicon would cool it down. The effort produced the 2N696, and Fairchild Semiconductor announced the product at Wescon 1958. The company found a large market, especially for military applications, and was only limited in sales by the amount it could produce. By 1959 the company had a whole family of products and the technology to build clusters of transistors into integrated circuits (IC).

ENCOURAGING THE VALLEY'S R&D

"Sometime in 1956, Fred Terman, John Linvill and Bill Shockley began discussing a path that Stanford could follow to create a laboratory in which graduate students would be able to do important research on semiconductor devices and technology," said James Gibbons, Reid Weaver Dennis Professor of Electrical Engineering, Department of Electrical Engineering at Stanford, and Dean of the School of Engineering from 1984-1996. Gibbons was at the center of the Valley's technology when The Valley of the Heart's Delight shifted to a valley built on silicon. And as with anything made from sand, high tides, low tides and the elements are bound to cause changes.

"They wanted to add new faculty in areas in which there appeared to be a very promising future," said Gibbons about the beginnings of the program he was destined to head. "They were interested in the possibility that university research on semiconductors would be such an area. To put this idea in context, it is important to note that, prior to the advent of transistor, university research

James Gibbons
Photo by Sally Richards

in electronic circuit design was concentrated on applications in which one assumed that the active components would be vacuum tubes that were purchased from commercial manufacturers. It was too expensive to set up facilities at a university for making state-of-the-art vacuum tubes. Universities simply bought vacuum tubes from suppliers such as General Electric and RCA, and then did research on circuit design using these components.

"The advent of semiconductors, and silicon in particular, changed all that. There was now the possibility that a university could actually set up a laboratory in which both graduate students and undergraduates could build transistors and other semiconductor devices, eventually including integrated circuits, though integrated circuits had not been invented at that time. So Stanford decided to try it," said Gibbons of the reverse transfer of technology project on which Stanford was about to embark. "You hear a lot these days about technology transfer from universities to industry. Here you had a situation in which technology was being transferred from industry to the university. I was hired and immediately apprenticed to Bill Shockley, where I met Gene Kleiner, Gordon Moore, Bob Noyce and the remainder of the eight who left Shockley Semiconductor Laboratories to form Fairchild Semiconductor.

"In fact, I wasn't there too long before Bob Noyce came around and said 'we're going to be leaving and forming a new company, and we'd like to have you join us.'

"'No,' I said, 'I want to be a professor. I am here to learn how to build semiconductor devices so that I can start a laboratory and launch a semiconductor device research activity at the university.' What happened is that the Fairchild eight did what they said they were going to do; and so did I. It is perhaps the most expensive decision I ever made!" Gibbons said about turning down his opportunity to become the "Traitorous Ninth."

"One of the interesting things about our starting the laboratory was that people at other universities were reasonably sure that we would fail. They were convinced that graduate students at a university would never be able to compete with the research that is conducted in major industrial laboratories. But semiconductors proved to be different, as Terman, Linvill and Shockley had hoped they would be. We did succeed. We built a semiconductor research lab; and when the integrated circuit came along

Photo by Sally Richards

we were ready to build another laboratory that concentrated on integrated circuit research. More recently, this building [the Paul G. Allen Center for Integrated Systems] was built because we developed the integrated circuit knowledge and research expertise that were needed to design special purpose integrated circuits for particular computing applications. We were poised to do something that no one else could. Looking back, it is clear that something we started 40 years ago has now been through several metamorphoses. As the field grew, we were able to stay current. We were able to attract funding and to prove that we could consistently do research that would make a difference in the semiconductor, computing and information industries.

"Whatever the field of research, our goal is to bring students to the edge of the field, so that when they leave Stanford to take jobs at TI, or Intel, or Bell Labs or other places, they will be at the leading edge of their field of research. For most of the last four decades, graduate students have moved from Stanford to very productive careers in modern electronics at a time when the demand has been rising rapidly. During this period we have also taken advantage of the opportunity to interact with Silicon Valley as it grew. Essentially, the Stanford solid-state lab and Silicon Valley grew together. One sometimes hears that Stanford started Silicon Valley. For the semiconductor industry, that is hardly the case. What happened was, for quite different reasons, the Shockley Semiconductor Laboratories seeded both the Stanford Solid State Laboratory and the commercial firms in Silicon Valley that make semiconductor devices, starting with Fairchild.

"Gordon Moore used to give a talk he entitled, 'The Rise and Fall of Silicon Valley,' where he listed three factors he thought were critical in the rise of the Valley," Gibbons pointed out. "The first one was Bill Shockley — when he said that people naturally assumed that this was because Shockley was one of the co-inventors of the bipolar transistor. However, Gordon's view was different. He felt that Shockley's most important contribution to Silicon Valley was that he formed an unstable company. Fairchild spun out of Shockley, marking what is generally regarded as the first time in Silicon Valley's history where a team of people left a company to start a new one that was designed to compete directly with the parent firm. We take such things for granted today, but there was a time when it was not commonplace," Gibbons said, comparing it to today's market where it is not only common, it's practically inevitable.

Sand Dreams

Fairchild's young entrepreneurial team had a dream that sand would transform the industry. It seems only fitting that the idea of applying silicon in the manufacturing of semiconductors would be used on the West Coast, where the dreamers of other centuries had searched so hard to find California's golden-grained coastline. Those countless tiny bits of sand were worth more in the process of making semiconductors than all of the gold the Spaniards had ever dreamed of finding, or that the 49ers had.

As with most large corporations trying to manage creativity, Fairchild Semiconductor's mother ship felt it necessary to find someone to dictate creation, the unmanageable chaos in which even the muses thrive.

"None of us had heavy management experience and they [Fairchild] thought we should have a manager and they brought in someone and he was awful. After we saw him manage, we decided we could manage ourselves. There was no magic to it. We took over the management of the company ourselves and Bob Noyce became the manager and Gordon Moore became the director of research. We felt we didn't need a manager and we were correct.

"They [Fairchild Camera and Instruments] were not used to stock options and incentives. Even at that time companies gave stock options to their key employees, but they didn't, so some people wanted to leave. Fairchild Semiconductor became a fairly large company," said Kleiner of the company's success. "We made more than 100 percent of the profit for the parent company. In other words, the other divisions were losing money. Something like this doesn't last forever. We owned 100 percent of the stock and they had the option to buy us out at three or five years at a predetermined price. They did exercise that option ahead of time." Each team member took their $300,000 from the buyout and applied it to their new dreams. They had developed the integrated circuit in 1958 and created a legacy.

"We were happy and they were happy. At that time I felt like a multimillionaire," said Kleiner of his first company sale. "I stayed at Fairchild a little longer until 61, and then started a teaching-machine company. We didn't have computers at that time, so this teaching machine was interactive with audiovisual presentation and it kept score. I think it was very ingenious. We used telephone type circuitry to control it; largely, group educational system for any age from relatively young children and industrial uses. We called it Edex, for Educational Excellence. We sold it to Raytheon, but they didn't know what to do with it, and didn't have any real plans when they bought it. No plans, no vision.

"It was a machine that controlled an answering box with five buttons. It lectured with audio and slides and pictures and asked students questions; it kind of goosed them to give answers to questions about the lecture," Kleiner said of the importance of interaction during teaching. "It wasn't so important that they answered the question correctly, it was important that they committed themselves to answering something. It was really concentrated thinking they had to do and every couple of minutes we'd ask them a question. I think it was a very effective educational device. We tried it out on some high school dropouts and tested them out on English grammar. Well, they couldn't have cared less about English grammar, but the machine got them so excited that they actually started to fight and say, 'I told you, you should have pressed that other button,' and things like that. We sold machines to Bank of America to train their loan officers on some basic things. It was a good machine and it worked. There are certain things a machine can do better than a teacher — a machine has infinite patience. Teachers weren't too happy about it. We used to say that any teacher who could be replaced by a machine, should be replaced by a machine."

Raytheon purchased the machine because it had language labs and wanted to combine the two. "They tried to do something with it, but they made it part of the missile division which at the time was a mistake because your buyers were people in the teaching industry and very sensitive to missiles," Kleiner said, shaking his head in disbelief of the marketing faux pas. "On my business card it said missile division, and that didn't really help me make sales. And then they tried to move me to Michigan City, Indiana," said Kleiner of the attempt to uproot him from Silicon Valley where his family was well established and happy under the Valley's waning noon sun. "Well, you don't move from Palo Alto to Indiana, and that was the end of it."

Kleiner would go on to co-found Kleiner Perkins Caufield & Byer, a venture capital firm that became legendary and an essential part of the Valley's history.

Brilliant Drinkers & Incredible Thinkers

While all of this incredibly creative inventing was going on all over the Valley, engineers and scientists and the followers they created gathered to debrief at places like Dinah's Shack, Clarke's, Walker's Wagon Wheel, The Cottage, The 101 Club and scores of other dark bars and hamburger joints. There they would unwind, talk about their latest frustrations and solutions, and make partnerships. It was a time of think tank-style creativeness, except it was done with a beer in one hand and another one on its way. It was common for these brilliant drinkers to scrawl chip designs on napkins and pass them around the table for discussion.

The Cottage
Photo by Sally Richards

Meanwhile, corporations and developers were purchasing land, and housing costs rose steadily. Schools were built to accommodate the rise in population and the economy was in good shape.

Noyce and Moore went on to found Intel and embraced Marcian E. "Ted" Hoff's microprocessor. In 1968 Hoff joined Intel and developed a chip small enough to fit into nearly any device, making it possible to dump the expensive, cumbersome computers the size of apartments that were being housed in chilled rooms inside corporations everywhere. Noyce found that a cluster of transistors and their connections could be successfully

The process of making silicon wafers
Intel

etched onto a single piece of silicon. Unfortunately, they were hardwired, meaning that these chips could only do the job they were designed to do. Hoff's breakthrough designed a set of chips that worked together to perform a device's function. Hoff knew designing a chip to run computer programs on its own would make the product very valuable. So it was that Hoff developed a chip to act as the Central Processing Unit (CPU). After some fine tuning, Intel launched the design of a general-purpose logic chip that could be programmed to take instructions. This meant that intelligence could be programmed by software and it no longer needed to be burned into hardware. This major breakthrough caused an evolution that changed the way America — and the world — worked and lived, and people everywhere started using calculators.

Hoff was named "one of America's seven most influential scientists since WWII" by the *Economist* and is credited with making computers available to the masses. There hadn't been an accomplishment like Hoff's since the creation of the Guttenberg press, a machine that brought print, thus literacy and knowledge, to everyone. Programmable intelligence kept shrinking and getting smarter and continues to do so to this day. This phenomena is a theory that Intel's co-founder puts into a nutshell called Moore's Law. This law, one that the Bay Area's economy seems to have thrived under, is described as the logic density of silicon integrated circuits increasing every year since the technology was created. The change affects price and size with each increment. Many believe the Valley's future success depends on the continuance of Moore's Law, while others aren't so sure.

As noon's sun dipped lower into the western sky, events were going on all over the world that would change the way the Valley's dreamers invented. An unfettered form of creativity was on its way, a generation of technology tempered by a devastating war that nearly tore the country apart and an unprecedented independence that set people freer than they had ever been before — free enough to invent a whole new world.

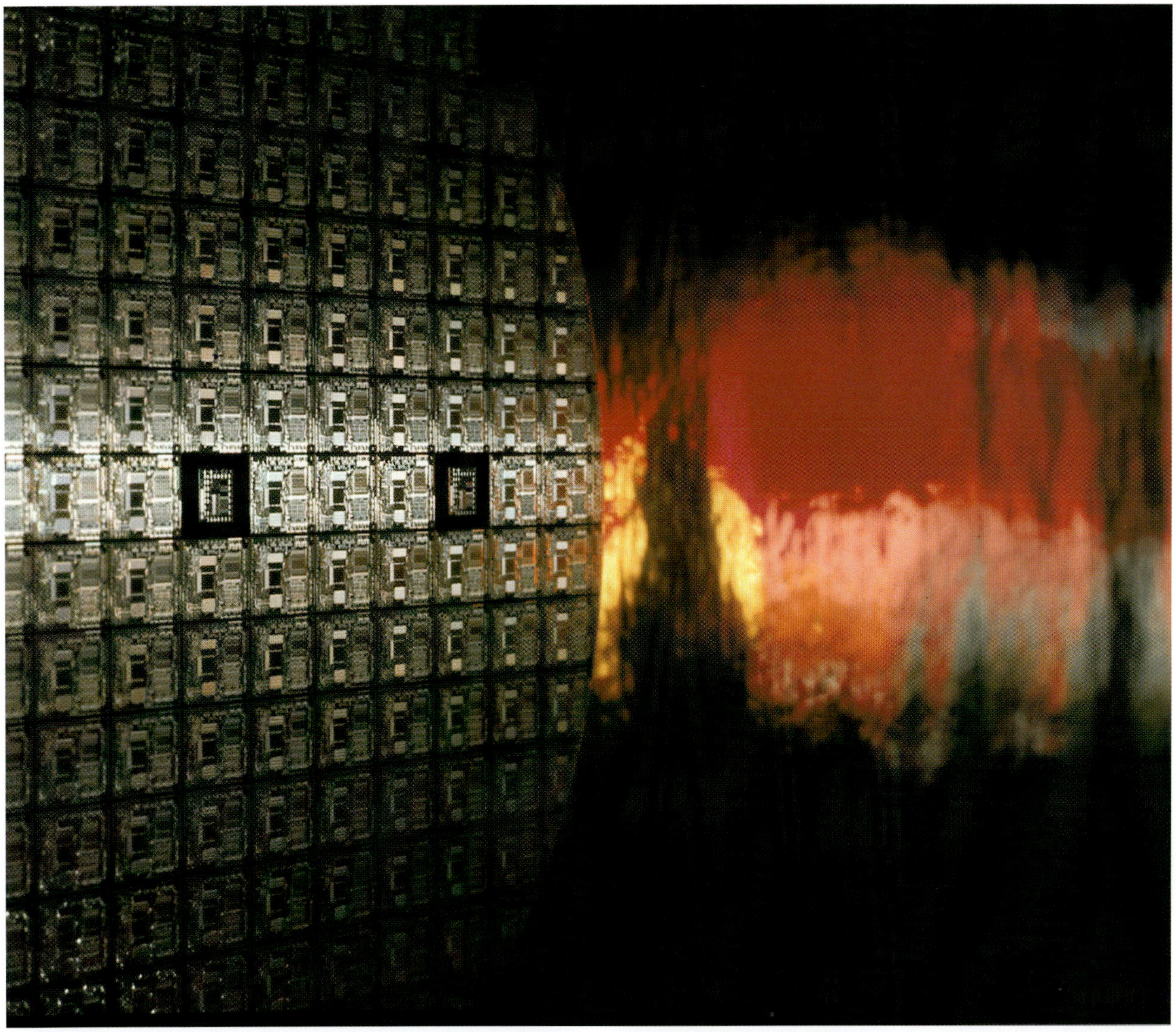

Intel

Andy Grove, Gordon Moore and Bob Noyce
Intel

85 | CHAPTER THREE

Photo by Sally Richards

afternoon

"We dare not forget today that **we are the heirs** of that first **revolution.** Let the word go forth from this **time and place...** to **friend** and **foe** alike... that the torch has been passed to a new **generation** of Americans born in this century. Tempered by war, **disciplined** by a **hard** and **bitter peace,** proud of our ancient heritage and unwilling to witness or permit the slow undoing of those human rights to which this **nation** has always been **committed** and to which we are committed today... at home and around the world. Let every nation know... whether it wishes us well or ill... that we shall **pay any price,** bear any burden, meet any hardship, support any friend, oppose any foe, to assure the survival and the success of liberty. This much we pledge... **and more."**

— John F. Kennedy's Inaugural Address, January 20, 1961

Photo by Sally Richards

CHAPTER FOUR

Afternoon in the Valley slipped rapidly toward the early twilight during the decades of the 60s and 70s. The era had a hard push from technology and all the excitement, greed and competition surrounding it. By the end of the 50s, Ampex had built a videotape recorder, the first transatlantic telephone call had been made via cable, the Soviet Union's Sputnik had sent signals from space, FORTRAN had become the first high-level language, stereo recording had been introduced, videotape was producing color images, data was being moved over regular phone circuits and Xerox had manufactured a plain-paper copier.

POLITICS AND THE PRESS

In 1962 Nixon ran for governor of California and young Stephen Wozniak, who would later found Apple, made his debut into Valley history. His mother, Margaret Wozniak, outgoing and outspoken, became president of the Republican Women in Sunnyvale (and would later become disillusioned

(Far right)
A photo from Ampex product archives shows their early equipment.
Redwood City Public Library Local History Room

Photo by Sally Richards

Harry Farrell and former President Richard Nixon
San Jose Mercury News

with the party and become an independent Democrat). Stephen and his two siblings were recruited for precinct work for Nixon. Stephen, always the engineering genius, used his amateur radio setup that he had built to help Nixon's campaign efforts and even ended up in a picture with Nixon on the cover of the *San Jose Mercury Magazine*.

"I have indeed lived through some exciting history," said Margaret Wozniak in a recent interview. "I watched the beginnings and subsequent growth of the Valley, and I am truly proud of the part my son played."

Political campaigns were fairly well covered in the local press, especially if there was a good hook or a rising star involved.

"I liked Joe Ridder," remembered Harry Farrell, reporter for the *Mercury* for more than 40 years and author of several books about San Jose's history.

"He was a big booster for city growth," said Farrell about one-time publisher of the *San Jose Mercury News*, Ridder. "When he was once asked what he thought about city growth and the orchards being replaced by homes, he replied, 'Prune trees don't buy newspapers.' When Nixon was running for governor in 1962, after losing to Kennedy in 1960, he came through here after some kind of meeting in Sunnyvale. He went over to Joe's house in the hills above Saratoga, overlooking the Valley, and I was there. We could see lights coming on all over the Valley, where 10 years earlier it would have been all dark. Joe happily told Nixon, 'and every light is a new subscription!' I drove Nixon from Joe's back to the St. Claire Hotel."

Farrell covered Nixon's campaign visits to the Valley and the memorable night protesters stoned Nixon in a flurry of angry rock throwing in 1970. "They threw rocks at his press bus and busted up windows and everything else. It was a big, big story. After I covered the 1962 gubernatorial campaign, he sent me a letter commending me on my fairness. After his downfall, he stayed completely out of sight for a long time, and then David Frost paid something like $600,000 for his first interview. I got about the second one didn't pay a nickel. I had him for about 25 minutes and he was talking about the books he was going to write. We weren't talking about anything important, and all of a sudden he said, 'You know that restaurant — O'Brien's [in San Jose]? I really liked that place; is it still there?'"

A Generation of New Voices

Politicians weren't the only ones making the news in the early 60s. Civil rights protests were raging throughout the country and changes were affecting the Bay Area and the world. People who had had no voices in society were heard for the first time. While the Valley had long been dominated by middle and upper-class, predominately white society, those living in their shadows — who worked in the fields and orchards, in the home and marketplace — finally stepped into the spotlight and spoke out for their rights.

Cesar Chavez was one of the Valley's early leaders for civil rights, and though his movement was spurred in support of the farm workers in his own town, his impact was national in scope.

Chavez, born in 1927, spent his early years on his family's farm near Yuma, Arizona. However, during the final years of the Depression his family lost its farm and was forced into migrant labor. From that point on, Chavez lived the life of a typical migrant child, wandering from farm to farm with the seasons. He recalled attending more than 65 different elementary schools.

As a teen-ager, Chavez and his family settled in San Jose where his father worked in a fruit orchard. Chavez dropped out of school after completing eighth grade so that he could help provide for his family. He spent many long years working the fields, but by the time he was in his 20s, he had tired of the labor. He was recruited by Fred Ross Jr. to work with the Community Service Organization (CSO) in East San Jose in the *Sal Si Puedes* ("Get Out if You Can") neighborhood. It was in his role as the national director for the CSO, which campaigned to register Hispanic voters, that Chavez became a skilled motivator and organizer. However, in the early 1960s when the CSO refused to reach out to the farm workers in Santa Clara Valley, Chavez set out for the San Joaquin Valley to begin his own rallying work.

Chavez settled in Delano to form the National Farm Workers Association (NFWA). When a band of Filipino farm workers in Delano pushed to create a farm workers union in 1965, Chavez joined the strike that eventually led to the creation of the United Farm Workers (UFW). The UFW strike and boycott of table grapes in the mid-1960s brought national attention to Chavez and his cause, making him perhaps the nation's first Mexican-American leader. In 1970, after five years and a successful nationwide boycott of grapes, the grape growers finally signed a union contract with the UFW.

Inspired by Gandhi and Martin Luther King Jr., Chavez used nonviolent acts of civil disobedience to advance his cause. He went on extended marches, walking hundreds of miles and gathering as many supporters. He fasted frequently to draw attention to his passion for the rights of farm workers. In 1968 he fasted 25 days to reaffirm his stance of nonviolence, and in 1988 he fasted 36 days in a "Fast for Life" protesting the pesticide poisoning of farm workers and their children. He drew support from President Kennedy and his family, California Gov. Jerry Brown and many other political and civic leaders.

Cesar Chavez outside the J.D. Martin Ranch near Tulare, California, c. 1965
Walter P. Reuther Library, Wayne State University

Although Chavez's tireless efforts were for the rights of all farm workers, his impact was especially strong in the Latino community. Through his leadership, Latinos found their voice in the community and were inspired to speak out not just through civic activism, but also through the arts, culture and politics.

In the same fields where Chavez fought for dignity for farm workers, two young children picked crops alongside their fathers. Though strangers then, these two would grow up to become teammates and Olympic athletes with a passion not unlike that of Chavez. Tommie Smith and Lee Evans would eventually meet at San Jose State University — both were members of the school's track and field team and sought an education to free them from a life of farm labor.

Smith came to San Jose State University in 1963 on a basketball/track/football scholarship, but it was sprinting that he chose to pursue. Evans transferred to San Jose State from San Jose City College in 1965. The two men became close friends, and as a result, their coach, Bud Winters, chose to place them in separate events. Smith's speed in the 200 and Evans' stamina in the 400 won them both world records in 1966.

Sprinter John Carlos joined Smith and Evans at San Jose State in 1967, having transferred from a Texas university where he had faced racism both on campus and in the community. While he found greater acceptance in San Jose, blacks were not free from injustices there.

In 1967 Harry Edwards, a former track athlete and a sociology professor who had helped recruit Carlos to San Jose, taught a class called Racial Minorities. He encouraged the blacks on campus, most of whom were athletes, to use their leverage to gain equal rights for themselves. Edwards and a group of students demanded an end to the housing discrimination on campus and fairness in all campus organizations, threatening to disrupt the first football game if necessary changes were not implemented. This local protest was just the impetus for a more powerful stance against racial injustice that would be spearheaded by Edwards.

Later that fall, he gathered a group of black athletes to consider boycotting the upcoming Olympics to protest racial injustice in the United States. Those who joined Edwards issued a list of demands to the U.S. Olympic Committee, among them that South Africa be barred from the Games. From this initial meeting the Olympic Project for Human Rights was born.

Smith lost his job at the Pontiac dealership in San Jose when his employer found out he was part of the Olympic Project for Human Rights. All the athletes supporting the project became subjects of hateful threats and discrimination.

Peter Norman, Tommie Smith and John Carlos accepting their medals at the 1968 Mexico City Olympics
John Carlos

Although the boycott was abandoned during the Olympic trials in Los Angeles, the San Jose State athletes could not ignore the cause — civil rights for all. Smith, Carlos and Evans all won their way from the San Jose State track to the Olympic track in Mexico City in 1968. As Smith and Carlos were led into the stadium with the other finalists just before their race, they agreed that if they won, they would do something on the victory stand that would make a statement to all those watching. Indeed, Smith won the gold and Carlos the bronze, and before the medal ceremony Smith handed Carlos a single black glove and told him what he planned to do. Overhearing Smith and Carlos, silver-medal winner Peter Norman of Australia accepted an Olympic Project for Human Rights button that Carlos offered to him, and he wore it during the ceremony.

The medals were hung around their necks and when the *Star Spangled Banner* began, the three athletes turned to face the flags. Then Smith and Carlos bowed their heads and lifted their gloved fists into the air. They wore no shoes, representing black poverty, and Smith's black scarf and Carlos' beads signified black lynchings. Their fists claimed victory for black power and unity.

Similarly, when Evans won the gold in the 400, he and his teammates, who had been warned by Olympic officials that a demonstration would ruin their future in athletics, made their own statement by wearing black berets, which they took off and waved during the national anthem.

The silent but heroic stance of Smith and Carlos resulted in their ejection from the Olympic Village and suspension from participation in any future Olympic Games. Although the men faced ridicule, persecution and even death threats, their gesture directed America's attention to racism in athletics and the nation as a whole.

A War that Never Ended

Conflict with North Vietnam, which had begun in the 1950s and escalated over many years, turned to war as America and South Vietnam began espionage tactics and air attacks against North Vietnam. This behavior continued for years. Eventually, so many American soldiers were killed in the rice fields and jungles of Vietnam that the government instituted a draft to replenish its armies.

In the meantime, middle-class America went on as it always had, some barely noticing the wars playing out on

Stanford University demonstrators, April 14, 1970
Photo by Gene Tupper, Palo Alto Times Collection History San José

Photo by Reynolds Crutchfield

their televisions. Teenage boys all over America lusted after Barbara Eden of "I Dream of Jeanie" and were thoroughly fascinated with Gene Roddenberry's futuristic world of William Shatner's "Star Trek." Ironically, both Eden and Shatner would later cash in on their pop-icon days, becoming Internet spokespeople in 2000. Grainy Vietnam war and civil rights clips were fed into America's televisions while Americans ate soggy fried chicken TV dinners on TV trays. This generation would make an easy transition from sitting in front of the television to sitting in front of computers, and in less than 10 years the latter would become an ingrained behavior.

By 1968 organized protesters all over America, especially those attending college, had answered the call to protest war. That same year Johnson announced he would not be seeking re-election. Nixon was voted into office claiming he had a "secret plan" that would end the war.

One of the many local protests against the war in Vietnam
Photo by Reynolds Crutchfield

The era that began with peace and love as its vision statement — a time when peaceful protestors met armed National Guard soldiers by placing flowers in gun barrels — had ended abruptly and left piles of human wreckage in its wake. Would new generations take away any lessons from the violence? They would certainly know how to manipulate their new best friend — television. At that time, peace protesters were putting the pressure on government officials to end the war as the violence that countered them was broadcast on national television. Campuses around the Bay Area held daily anti-war, anti-Nixon protests that often turned ugly when police were called in.

Americans were hit with the oil embargo and faced petrol rationing and long lines that wrapped around entire blocks. The days when one could jump in the car and drive across the country with ease had ended. Americans wondered if there would be enough resources to see them through the next year. It was apparent that the gluttonous consumption of natural resources after World War II had come to an end. Conservation became a trend that would become a political part of California's future.

TECHNOLOGY CONTINUES UNINTERRUPTED

During the pre-Vietnam protests and throughout the war, innovations in the computer industry continued with hardly a hitch. The Academy of Motion Picture Arts and Sciences presented Ampex with an award for technical achievement in 1960. Throughout the early 60s, there were also a number of commercially viable computer manufacturers making their way into businesses, universities and government markets. The computer companies leading the pack included IBM, Burroughs, Honeywell, Control Data and Sperry-Rand. The computers used solid-state technology, and transistors instead of vacuum tubes. The most important model was IBM's model 1401, whose popularity created a loyal following of IBM customers.

A highlight in technology occurred in 1964 when John Kemeny and Thomas Kurtz created BASIC (Beginners All-purpose Symbolic Instruction Code) programming language at Dartmouth College. In 1967 IBM built the first floppy disk and Big Blue continued to work harder to stay ahead of the stiff competition it would inevitably face.

One year later the Air Force made a decision, although it didn't know it at the time, to gift the world with the most powerful tool of communication ever invented. The U.S. Air Force contracted Paul Baran of the RAND Corporation to develop an electronic backbone communications system that would survive a nuclear attack. The backbone would have no central hub; therefore it would not be broken if a nuclear bomb annihilated one of the locations linking the system. The Defense Department's Advanced Research Projects Agency (DDARPA) headed the project that linked scientists via a network at UCLA, Stanford, UCSB and the University of Utah. This project, dubbed ARPANET, became what is known today as the Internet. It's ironic that such a powerful communications tool originated during the Cold War by a government that, several decades later, is trying to regulate it.

Xerox, the New York-based company founded in 1903, had rescued the Halliod patent for dry-ink plain-paper copying technology and soon every business in America had to own one of its copy machines. The result of the company's wealth was the formation of Xerox PARC (Palo Alto Research Center) in 1970, a place where brilliant minds from all over the world created technology in a well-funded environment.

PARC thrived during the 70s, and developments such as GUI (Graphical User Interface) were completed there with the invent of Alto, a computer that Xerox saw as unmarketable. This machine would later help make a revolution. Although PARC was known as a laboratory where little research was capitalized upon in the 70s, today's PARC is quite a different place with many profitable startups in its portfolio.

Survivors of War

At 4:30 a.m. on April 30, 1975, two Americans were attacked by rockets at Saigon's Tan Son Nhut Airport where they were guarding the U.S. Embassy's lift off. They were the last Americans officially killed in the Vietnam War. Only hours later, North Vietnamese tanks rolled into Saigon and ended the bloody battles that had lasted 15 years and taken the lives of nearly 2 million soldiers and guerillas from both sides and more than 200,000 civilians.

Free of another war, American soldiers returned home to be met with protest instead of ticker tape. Anti-war sentiment and news of drug use and violence against civilians had been brewing long before their homecoming. For many, reintegration would be painful, and for others it would be impossible. Some would spend their entire lives at the veteran's hospital in Palo Alto made popular by Merry Prankster Ken Kesey, a La Honda resident, in his book *One Flew Over the Cuckoo's Nest*. A large number of soldiers stationed at local bases were discharged and joined the burgeoning high-tech industry and went on to successful careers. For others, the war may never end, and we see casualties of the Vietnam era taking up residence under the Guadalupe Bridge and highway overpasses throughout the country.

Another group of survivors came to Silicon Valley from Vietnam following the war. These men, women and

The 10th anniversary of the end of the Vietnam War was commemorated in St. James Park in San Jose on April 28, 1985.
Photo by Norbert Von Der Groeben, Palo Alto Times Collection History San José

A Vietnam vet and his daughter leave the office of American Council on Nationality Service with their 20 pounds of rice.
Photo by Joe Melena, Palo Alto Times Collection History San José

children were not returning to their homeland, but rather fleeing from it. Vietnamese refugees flooded American shores in 1975, the beginning of a wave that would change the very face of the region.

The first Vietnamese immigrants to the United States were processed at Camp Pendleton in Southern California, at Fort Smith in Arkansas, and at Fort Indiantown Gap in Pennsylvania. The new arrivals were then scattered across all 50 states in the government's attempt to avoid impacting any one area too heavily. But what followed this systematic placement became known as the "second migration." Many Vietnamese chose not to stay where they had been placed and instead flocked to a few regions — one of which was Silicon Valley. The California Bay Area offered a climate similar to that of their homeland in Vietnam, and jobs were plentiful.

The booming computer industry, and later the semiconductor industry, provided assembly and labor jobs that didn't require English skills. In computer manufacturing facilities, it was not uncommon to hear Vietnamese and Indochinese spoken instead of English. There were plenty of opportunities to work overtime and the Vietnamese became known for their Puritan work ethic.

Word of mouth of the Valley's opportunities spread among the Vietnamese populations throughout the United States and also in their homeland, and a Vietnamese community soon developed in San Jose. At first just a few restaurants and shops, the Vietnamese community quickly flourished, filling in vacant buildings and entire shopping centers. The 1980 census reported 11,700 Vietnamese living in Santa Clara County; by late 1982 that number had escalated to 40,000 or more. It was this influx of Vietnamese refugees that initiated today's strong Asian-American presence in the Valley and the growing acceptance of a diverse community.

New Faces in Politics

The effects of the civil rights movement during the 1960s, which empowered individuals and entire communities, began to surface in Silicon Valley's political scene during the 1970s.

Norman Mineta was elected to the San Jose City Council in 1969 and was the first Japanese-American to serve as a council member. In 1971 he replaced Ron James as mayor, and in 1974 he was elected to Congress.

In 2000 President Clinton completed this remarkable career by appointing Mineta as Secretary of Commerce, the first Japanese-American and the first native-born Californian to hold a cabinet post. Mineta's emerging role in politics was especially significant because he became a representative of not only his congressional district but also the Japanese-American community, many of whom had grown up during World War II.

Mineta's father had immigrated to the United States from Japan in 1902 as a 14-year-old who spoke no English and knew no one in America. By 1920 he had started his own insurance business, which his son Norman eventually took over. Though by law he could never become a U.S. citizen, he worked hard for his family's future success, providing for his five children and putting them all through college. When World War II began, the Mineta Insurance Agency was a successful business that supported the entire family.

But by 1942, young Mineta had witnessed the suspension of his father's business license, the arrests of honorable members of his Japanese neighborhood, and the institution of a curfew and travel restrictions. When President Franklin Roosevelt signed Executive Order 9066 on February 19, 1942, which proclaimed that all people of Japanese ancestry be moved from the West

Norm Mineta, 1976
History San José

Coast because of "military necessity," the family's savings account was confiscated and their dog, Skippy, a wire-haired terrier, was given away.

Mineta wore his Cub Scout uniform and carried along his catcher's mitt and baseball bat the day he and his family were evacuated. However, when he reached the train station, he had to leave behind his bat — it was considered a weapon. Mineta and his family were transported to Santa Anita racetrack in guarded trains with all the shades drawn. After three months there, they were moved to an internment camp at Heart Mountain, Wyoming, a more permanent settlement. Wyoming was bitterly cold and the transplanted Californians had little to protect themselves from the elements.

In 1943, members of the Mineta family were given permission to leave — first his father, who went to Chicago to teach American soldiers Japanese, then Norman and his mother.

Mineta was one of 120,000 Americans of Japanese descent who endured the internment camps during World War II. While a sense of shame hung over the Japanese community, some, like Mineta, chose to rise above the tragedy they had endured. Mineta's career in civic leadership was born out of his purposeful decision to prevent future tragedies from destroying innocent people's lives.

During Mineta's 21 years in Congress, he was known for transforming Silicon Valley's transportation system and creating the Congressional Asian Pacific Caucus and Research Group, which provides awareness of Asian American issues. However, perhaps his greatest achievement was obtaining the Civil Liberties Act of 1988. This law offered a long-overdue apology and redress to Japanese-Americans who had suffered the forced internment during World War II. The day Mineta signed the Civil Liberties Act, he, in essence, lifted the shame that had hung over his family and the Japanese-American community for decades. Mineta's rise from his Japantown boyhood to war internee to Congressional leader helped redefine the faces shaping the Valley's politics in 1970s.

While Mineta was making strides for Japanese-Americans, Janet Gray Hayes soon emerged on the political scene as a new voice for women in politics. Elected mayor of San Jose following Mineta, Hayes was the first woman to hold the mayorship in a large American city. During her two terms lasting from 1974 to 1982, women held a majority of the seats on the San Jose City Council and board of supervisors for the first time. Hayes claimed that the Valley became the "feminist capital of the world" during her leadership.

During Hayes' second term, San Jose municipal workers went on strike demanding "comparable pay for comparable work." In the late 70s, the union, Local 101 of AFSCME, had pressured the city to hire a consulting firm to evaluate and rank city jobs. Based upon the firm's research and evaluations, it was recommended that the city give pay raises to women in many positions. While the recommended salary increases were implemented in 330 managerial positions, the city claimed it could not afford salary raises for secretarial and lower-ranking jobs. It was this decision that spurred the nation's first strike over pay equity in June 1981.

The strike continued for 10 days and, despite resistance from the council, ended in a two-year contract providing $1.4 million toward comparable worth and an 8 percent cost-of-living raise. For the most part, San Jose's municipal workers received what they fought for — pay increases

A supporter stands on the sidewalk off 1st Street in San Jose while listening to speakers striking for "comparable pay for comparable work."
Photo by Susie Gow, Palo Alto Times Collection History San José

averaged 17.6 percent. The success of the San Jose municipal strikers in 1981 reinforced the region's claim as a feminist capital and was a catalyst for other groups nationwide to fight against wage discrimination.

THE EVOLUTION OF REVOLUTION

While people were endowed with new freedoms as a result of the social changes occurring in the Valley and the nation as a whole, they also looked toward technology for personal empowerment. There were minicomputer innovations in the 70s, including the Intel 4004 chip in 1971. This microprocessor met the needs of many applications. The new design made consumer products such as televisions, microwave ovens and electronic fuel injection systems affordable to the general public. Digital Equipment Corporation's minicomputer hit the market but only academic and research markets seemed to be responding well, leaving the personal computer target wide open.

"I headed west because I wanted to be at the center of things and I always felt it was important to be where the decisions were being made," said Nolan Bushnell, founder of Atari, the computer company that started the billion-dollar computer game industry. "I had a strange dilemma when I graduated from college with an engineering degree. To take a job as an engineer I would have to actually take a pay cut."

When Bushnell was 15, his father died unexpectedly, which forced him into the position of running, operating and eventually, after all outstanding contracts were completed, closing down his father's construction business. He worked his way through college as the manager of the games department of an amusement park that had a higher game income per capita than any other amusement park.

"I had several very good offers to stay in the amusement park business — significantly more money than a beginning engineer," said Nolan of his decision. "I felt that I could always go back, my reputation would stay known for three or four years. If you let a fresh engineering degree go fallow it becomes nothing, and I'd always been interested in technology. I'd already done the amusement park thing, so I decided to come to California and pursue engineering. This is where things were happening, this is where the semiconductor boom was taking hold. There was the crackle of new opportunity here, but it was more like a technical opportunity. From an engineering opportunity you kind of wanted to be where all the neat stuff was — it was more like that than making a financial killing.

"When I drove here in my 1958 Chevy, I found a bunch of prune orchards and the start of the technical industry — the concrete tilt-ups were just starting. For the first six years I was here, if you owned a chainsaw you would always have firewood for your fireplace. As they were tearing out the prune orchards to build the high-tech buildings, they'd put a sign that said "free firewood." It was an exciting time, but it really hadn't gotten into full gear. It was still kind of quiet and laid back but you knew there was neat technology happening."

Bushnell got a job as an associate engineer working on VCRs with Ampex, a company that developed Instavision with Toshiba. Also entering the VCR race in the early 70s were N.V. Philips, which introduced its own videocassette recorder format in Europe, and AVCO, which released a solid state compact Cartrivision. Bushnell was walking into a quickly developing market.

"I was out in a division called Video File in Sunnyvale. Ampex was based in Redwood City. I was just an underling; Ampex had several thousand employees. I

Nolan Bushnell
Photo by Sally Richards

Photo by Sally Richards

was not with the movers and shakers at that time. In the back of my mind I always felt I was an entrepreneur. I had a company of my own when I was in college and always felt it was something I wanted to do. I always looked at it like if I was being paid to do something, then there was someone else paying me to do it and they were making more money off my labor than I was getting paid. I always looked at it as selling your labor wholesale rather than retail.

"I actually started thinking games in college. I was an engineering student and this is when video terminals were just starting to get connected to regular computers. You'd see what the economics were at the amusement park and then you'd see the stuff in the lab at the university and think, 'If I could play games on the video terminals and put them in the amusement park, they'd make money.' Then you'd look at the big mainframe costs and you'd divide 25 cents into $7 million and the math didn't work. I just felt there would be a time when the economics were right. Costs were getting cheaper and the minicomputer was actually the thing that started me back on the road to building a video game. I actually designed a system that was just on the edge of being economical but the mini-computers were still too slow and video is very demanding. It takes a lot of data flow and when you're dealing with computers with 500 kilohertz cycle time — it's really bad. A lot of people don't realize that the first video games did not use microprocessors. Microprocessors really didn't get any kind of real power until 1976, and this was 1970, six years before the microprocessor came along. The early games we developed didn't really execute programs per se; they were fancy signal generators. I know it sounds crazy, but if you wanted to put a digital score up, you actually put a four-bit set of flip flops to count the score, decode them and display them on the screen. It wasn't like you just decided to put something on the screen. It was actually a chip, part of the hardware design."

One of the precursor games to Bushnell's Pong (sans microprocessor) was developed during the Cold War by a noteworthy inventor who had developed an electronic table tennis game in 1958. His name was William Higinbotham and he was in charge of instrumentation design at Brookhaven National Laboratory (BNL), a government-supported nuclear research facility in New York. Higinbotham had his roots in radar design and designed timing devices for the Manhattan Project. BNL was implementing a PR campaign to keep locals from being suspicious of their work in the lab. In order to make things more friendly and fun in the sterile environment, group tours were set up for civilians and Higinbotham wired up some entertainment: an oscilloscope where a net and court were displayed in rapid alternation, with a ball motion on the screen. The game, without the micro-processor, was extremely slow in comparison to the one Bushnell would develop more than a decade later.

While working his day job at Ampex, Bushnell toiled in his garage nights and weekends until it got too cold, and then he moved his two daughters into one bedroom and set up a lab in the other. Bushnell, the original "Game Boy" of Silicon Valley, knew there was money in electronic games — he just had to figure out how to make a market.

"I think the very first time that I saw the little rocket ship on the screen and I could move it, I thought, 'This is too good, it looks too neat.' Three months later I did have a product. I went to a company called Nutting Associates to manufacture and market it. They had done a little thing called Computer Quiz, which was nothing more than a coin-operated slide projector that gave you a score. They put it in bars and grocery stores. Basically what I did was license the technology to them, and they said, 'We don't know anything about this newfangled stuff, why don't you come as chief engineer and get the thing into production.' So, I quit my job at Ampex, went to work for Nutting and put the product into production.

(All photos this page)
1984 Apple campaign
TBWA/Chiat/Day

"What happens — and the wonderful thing about Silicon Valley — is that it's impossible to work here without sitting next to somebody who you perceive to be an idiot and who gets very rich. You say, 'How can this be happening? This guy's an idiot and he just made a million dollars, and I'm sitting here, so who's the idiot?' What really happened with Nutting is that I could see all the mistakes they were making with my baby and it really bugged me, so I decided to license it to someone else. I did another deal with a company in Chicago and moved out into my own contract design house called Atari [meaning checkmate in the board game, Go]. I had originally sold a driving game to the company that was funding my research, but the very first game we came up with was Pong. It was sort of a training exercise for Al Alcorn, the first engineer that I hired. I just defined this silly game, but once he got it mocked up it was a lot of fun. So I thought, 'Maybe we could get this company to take Pong instead of the driving game.' They didn't; they said it was too simple and it was a two-person game. They wanted the driving game. All of a sudden we had Pong, and it was ours." Bushnell and his team built the Pong arcade game and put the prototype of the game in Andy Caps Tavern in Sunnyvale, a place where the Valley's engineers hung out.

"The machine had been on location for three days and we got a service call," recalled Bushnell. "We thought, 'Oh no, it's buggy, it's not robust enough for that environment.' What had happened was that it had been played so much that the quarters had completely filled the coin box mechanism and it just couldn't hold any more quarters. We thought, 'Now there's a problem we can fix.' We knew we had something, so we got all the money we could beg, borrow and steal — so that we could build 11 machines. So we ordered the parts, built them, shipped them and never looked back. We started with 11 machines, built 26 machines, collected the money for those and so on. We built the company one machine at a time, operating in a positive cash flow, and figured out just-in-time inventory before it was a buzz word."

Atari went on to develop Tank, Breakout, Asteroids, Battlezone and Centipede, the games that set the pace for the gaming industry. Both Steve Jobs and Stephen Wozniak, founders of Apple, worked at Atari for a time. During that period, Jobs showed Bushnell his plans for the home computer, offering him a third of Apple for $50,000. Unfortunately, at the time, Bushnell's funds were tied up in his next startup, Chuck E. Cheese Pizza

Time Theater. However, he saw Apple's potential and referred Jobs to venture capitalist Don Valentine. In 1975 Bushnell sold Atari Corporation to Warner Communications for the sum of $28 million and went on to make the covers of magazines and newspapers nationwide. Twenty-five years later, Bushnell would move to Los Angeles and build an Internet startup called uWink.com Inc. to once again revolutionize the gaming industry.

Years after Bushnell sold Atari, the company's logo appeared many times in the props of *Blade Runner*, a Warner film, which was a futuristic blockbuster about Los Angeles based on a book by Philip K. Dick. It was directed by Ridley Scott who would later direct Apple's award-wining "1984" commercial.

Jobs and Wozniak made sacrifices to start Apple — Jobs sold his VW microbus and Wozniak hocked his HP scientific calculator — and the two built the Apple I in Jobs' garage. Wozniak, willing to become a full-fledged entrepreneur, quit his day job at Hewlett-Packard to become VP of R&D.

It was 1979 when Jobs took his legendary stroll through PARC and liked what he saw in GUI. Several of PARC's code jockeys shared Jobs' vision and joined Apple. Valentine had introduced Jobs and Wozniak to Kingmaker Mike Markkula (a former marketing manager at Intel) who had the savvy to bring Apple to fruition.

In 1984 Apple made marketing history with its ad for Macintosh in its "1984" campaign. In contrast with today's commercials that are shown until the viewing public's eyes bleed from repetition, the commercial was aired only twice. The first airing was on December 15, 1983, on a remote Idaho station in order to qualify for that year's advertising awards; the second was during Super Bowl XVIII. The commercial won 35 awards, including the Grand Prix awarded at Cannes, for Apple's advertising firm, Chiat/Day.

The commercial, based on George Orwell's concept of a populace controlled by the government, had the tag line, "You will see why 1984 won't be like 1984." In a sense, the commercial epitomized the shift from the Cold War era to the free thinking that went on in the late 50s and 60s and led to the creation of the personal computer in the 70s, a tool of evolution — revolution.

The Macintosh freed users from the need to type in commands and gave them the ability to control the cursor with a mouse. Computing was now easy enough for anyone to learn, and the world would never be the same. IBM's "Little Tramp" ads, using a dead silent screen icon, couldn't hold a candle to the stir Apple had created. Ironically, Apple, a company that began using such innovative marketing ideas, shifted to using dead icons of history to push product in its "Think Different" campaign in the 1990s.

One of the early Apple converts was Guy Kawasaki, who became the company's evangelist and managed relationships between Apple and third-party developers as well as Apple-labeled software products. Kawasaki, a Stanford graduate, had moved to Southern California when he was offered the job. "The big picture is that when I saw the Macintosh for the first time — that was it. I fell in love with the product in the first five minutes — which is how long it takes for an evangelist to get a product. I knew it was a winner five minutes after the demo started," said Kawasaki who led the "1984" campaign.

"Word processing changed my life. It was so much better than a typewriter that the scales fell off my eyes." What was in the air when Kawasaki arrived in Cupertino,

Guy Kawasaki
Photo by Sally Richards

home of Apple? "Total optimism that we were going to make the world a better place with Macintosh and send IBM back to the typewriter business holding its Selectric balls." Kawasaki would continue revolutionizing technology into the 90s when he founded Garage.com, a venture capital company with a computerized infrastructure and more than 700 partners.

The battle between Apple and the PCs still rages more than two decades later. Other manufacturers of home computers would come and go in the Valley, including Atari, and the market would become increasingly fragmented as Asia joined in the competition in the 80s.

PARTNERING FOR THE FUTURE

Apple continued to make valuable relationships and bring believers to the table. One of those partnerships was with John Warnock, co-founder and CEO of Adobe, the world's third-largest software company — and former PARC employee.

"It was the world's greatest sandbox, well funded and with some of the world's best computer scientists," said Warnock of PARC. "I went to PARC because it had an incredible reputation and there was obviously very interesting stuff going on there. After being there, I had hoped there was some way to incorporate the technology that was being developed into product — to get it out there. Chuck [Geschke, president and chairman of Adobe] and I spent two years trying to do that. We had a very hard time getting any of the technology transferred into product. So we got frustrated and left. It's sort of funny: the 914 copier in the 1969 through 1970 timeframe was Xerox's money machine, this was really the home run — they started generating cash like crazy. The company grew rapidly and then they said 'Gee, what's our future going to be like?' The people who ran the company sort of didn't know the answer to that question, so they founded PARC. But there was no natural communication between Xerox the corporation and its research arm.

"The people running it [PARC] believed that it was those guys' [Xerox] responsibility to come and get stuff, but they were doing so well with their machine that they weren't inclined to go get stuff. Lack of communication. There were valiant attempts to build the communication path, but it was like having two bodies and both were growing but they were never symbiotic.

"I walked into Chuck's office one day and said, 'We could sit in these really cushy jobs for another 10 years, or we could actually go and try and do something.'" At the time Warnock founded Adobe, Sun had started up a year before and Microsoft was still a small company.

"Rather than sit around and talk about it, I decided to fly to Salt Lake City to talk to my old thesis adviser, tell him what we had in mind and see what he thought. I did, and he immediately introduced me to Bill Hambricht [Hambricht & Quist] and he provided the first round — two pieces for the total of $2.2 million. It lasted over two years. We were profitable when we struck a deal with Apple. The contract brought in a prepayment of $1.5 million. Apple then invested in 20 percent of the company and that gave us a bunch of money — it was all that we needed. There were lots of startups founded around the time of the depression in the early 80s. There were five or six that were starting in publishing, not many survived.

"It [Adobe] almost always seemed bigger than we ever thought it was going to be," said Warnock of the company's growth. "We doubled every year for the first three or four years. And then we had pretty stiff competition. HP was

John Warnock
Photo by Sally Richards

Disappearing one by one to make way for commercial buildings, by 1979 the drive-in was a thing of the past in Silicon Valley.
Photo by Joe Melena

doing its own thing when IBM signed up with us. HP gave up and became a customer. At that point, when we had the world's largest printer manufacturer as a customer, we sort of knew we had a very strong share in the market. We had just started with the applications business in 1986, and Illustrator was the first of our applications. We thought you had to be pretty big to be in the applications business and critical mass was probably $10 million a year. We got to 10 and then reached 25 — postscript was funding everything — and still weren't making any money. So then we thought, 'Maybe it's bigger than that,' so when we got to $50-75 million a year, we really started to make money. People who think you can get into the software business and make money by selling packages through distribution... it's tough."

The 80s were tough on the Valley and many companies did not live to see the end of the decade. Software and networking became the up-and-coming industries in the area and the Valley was soon growing strong again. As the late afternoon sun began to set on the 80s, the birthplace of technological revolution was poised for another change.

Adobe is located in downtown San Jose.
Photo by Sally Richards

Photo by Sally Richards

 twilight

"In the realm of scientific observation, luck is granted only to those who are prepared."
— *Louis Pasteur*

CHAP

Photo by Sally Richards

CHAPTER FIVE

As the sun slowly set, dipping into the twilight of the last millennium, many probably stood back objectively and felt that they had lived quite a few lifetimes in this place of reinvention. The Valley has been through many metamorphoses and accomplished things that no other civilization on earth has achieved. We now stand on the eve of a new dawn wondering what we are capable of accomplishing next.

As the days passed, so changed our traditions, even the way many generations of Valley youth were raised. By the time kids hit high school in the late 70s, they weren't spending their idle summertime hours in the orchards picking fruit; by then many of the orchards had gone the way of the chainsaw. Instead, many grew up with both parents working and became what society calls "latchkey kids." They grew up independent and spent their summers on their parents' computers, learning code, playing video games and having no patience for anything outside of their span of attention — which over the years of instant everything was becoming shorter and shorter.

One of the changes reflecting the sophistication of Silicon Valley's youth was the closure of the Valley's theme parks — Santa's Village and Dinosaur World in Scotts Valley and Frontier Village in San Jose.

(Far right)
Frontier Village's Wild West Drama at a mock high noon shootout
*Photo by Mike Roberts, Color Productions
History San José*

Marine World theme park packed up its animals and left for another area where land wasn't in such high demand. Oracle soon moved in to fill the Valley's needs.
Redwood City Public Library Local History Room

Lawrence Hollings designed many of the miniature golf courses and theme parks that helped shape a nation's dreams.
Photo by Sally Richards

"The people who build theme parks are bottom line," said the late Lawrence Hollings, San Jose resident and designer of Frontier Village and co-designer of Disney's It's a Small World ride, in an interview during the mid-80s. "They don't understand how people react to abstract ideas. They see the roller coaster — and that's all they want. Walt Disney used to say, 'There'll be a roller coaster in this park over my dead body.' Ten years after he died there were four. If I had my choice I'd build a fantasy park. I'll never do that until some kid who grew up on *Star Wars* decides that he wants to build a park — by then I'll be too old," Hollings laughed as he thought of the years it would take for the kid to grow up and get venture capital to build a theme park.

This was Hollings' response to Frontier Village closing. Once one of San Jose's main tourist attractions, the park was closed in the early 80s when the owners saw that Marriott's Great America would probably take away major revenue. The "white-knuckle parks," as Hollings described Great America, had replaced fantasy. The parks in this area before Great America were based on themes —

western, Santa, dinosaurs. The only survivor of that era is the Santa Cruz Beach & Boardwalk, which has a renowned wooden roller coaster, one of the best in the world, and the Pacific Ocean as a backdrop. American kids' tastes shifted radically after the invent of the video game, and after the showing of *Star Wars,* well they just weren't patient enough to imagine. Everything had to be instant — in real time and on the screen. And away the theme parks went.

Old and new stand side by side in San Jose's downtown. *Photos by Sally Richards*

"Frontier Village was three blocks away from my house," said Judy Stabile, deputy director of the San Jose Downtown Association. "It was a safe and delightful place that was perfect for elementary school-aged children, and a wonderful source for summer jobs in San Jose. It's sadly missed." Stabile was the driving force behind the Hayes Mansion (next to Frontier Village) project that turned the condemned mansion into a historical landmark. She was also responsible for finding the city dollars and the right corporate partnership to turn the mansion into the successful conference center it is today.

Stabile's main job currently is to renovate San Jose's Fountain Alley area (an area between San Fernando and Santa Clara streets and First and Second), where some of the last historic retail remnants of a bygone area can be found. In the 80s, growth caused San Jose to knock down eight square blocks of the city where many historical buildings were located. Recognizing the city has a great opportunity to preserve history, Stabile works hard with retail and entertainment companies, restaurants and landlords to make sure that San Jose does save its past. Stabile wants to save the Fountain Alley area from becoming an unrecognizable, what she calls, "city of *generica* suburbs."

"It won't be what it used to be because people aren't like they used to be," Stabile said of the new retail area where, with the competition from the malls, small clothing stores have no chance. "San Jose's retail attracted the entire county and you could get there by trolley — you bought your prom dress and christening outfits in those stores. That's never going to happen again. Downtown is emerging as a good-time place where there are retail specialty stores. A big addition to the area is the House of Blues, and the Fox will have Opera San Jose as the resident company and share it with other groups." Stabile's personal crusade to save this area is quickly coming to fruition as the panhandlers and troublemakers shift to other areas and the well-patrolled Fountain Alley District becomes a hot nightlife spot for Silicon Valley's young crowd who have lots of IPO money to spend. "There's something inviting about a handmade building — to imagine brick laid upon brick, to touch that handmade façade, it can really transport you back to another time."

Another woman trying to save San Jose's historic past is Alrie Middlebrook, president of Middlebrook Gardens. Middlebrook has a knack for bringing garden areas back to their turn-of-the-century splendor. Her talent is sculpting gardens, and one particular San Jose project she is involved in has her planting the same heirloom plants one would find if visiting the area in the late 1800s and early 1900s. The project is a partnership with some of San Jose's highest-profile developers. The area is a small neighborhood near Guadalupe Expressway and downtown. The property has been designated for a small specialty shopping enclave with turn-of-the century ambiance where shoppers can go from house to house to browse or dine. Currently, many of the Victorians that have been relocated there to make way for urban sprawl look like they've seen better days. To look at the area now, one sees rundown wooden homes up on blocks, some with plants growing from their roofs and a grassless dirt beneath them. The area is surrounded by a chain-link fence with padlocks. It looks forgotten, but Middlebrook has high hopes for the project that she wishes to work her creative magic upon.

"Italian immigrants came to California for the Gold Rush, and after the rush they came to the Santa Clara Valley and bought land. It was the beginning of a period of time where there was a large French and Italian immigration," said Middlebrook of the project. "The center of this community in San Jose during the 1880s and 1920s was the church and the Hi-Life Hotel. My task for this project is to re-create the back and front yards as they

Middlebrook Gardens in San Jose during renovation
Photo by Sally Richards

Mike S. Malone stands in front of his historic Murphy House in Sunnyvale.
Photo by Sally Richards

would have been during that time. I went to Italy and found quite authentic gardens, trellising and objects from that period. I even found heirloom flowers and vegetables from that time. I have struggled for the last five years to keep this project alive. Other projects are affecting ours and we're still struggling to get financing."

Middlebrook feels confident that the project will one day see completion. Until then, she is very busy working on other projects such as the Mexican Heritage Garden and the Chip Garden, no doubt an homage to the technology that has taken over Santa Clara Valley. Middlebrook's art form is to create a place where people can relate to the earth as it once used to be, to introduce its historic past to present time. "The real thing about these projects is that I want to introduce people of the Valley to the link of the land through the gardens and homes. We live on the earth, we build houses on the land — but we live on the earth, and I love the opportunity to create a garden that helps people understand that."

Walking Among Giants — the Future Meets the Past

One man who understands his relationship to the ground underneath his feet is Michael S. Malone, author of *Microprocessor: A Biography* and editor of *Forbes ASAP*. "It's like Troy, there's no one Troy, just six different layers of Troy spilled on the debris left from the previous one. You start with Bill and Dave in a garage and the Traitorous Eight and all the way back to de Forest and his laboratory — these guys, and the legacy of them, may be to establish a worldwide revolution. This place is famous for having no history, but these guys were like giants striding the earth. Who would have thought that just because David Packard decides that you can come late in the morning — he established flexy-time and stock options — that it would actually transform the world's economy 60 years later? But that's exactly what's happening.

(Both photos) Open space in Silicon Valley is replaced with homes and businesses for a growing population and industry. *Photos by Sally Richards*

When I was young, I saw them as giants, when in fact I really was walking around amongst giants. Their impact was even greater than I'd ever imagined.

"It used to go 'As goes New York, so goes America,' well now it's 'As goes Silicon Valley, so goes the world's industry.' You look around the world and you'll see the world imitating the Silicon Valley style; the suit is disappearing, cubicles are everywhere, the flattened organization — those factors are cropping up everywhere; it's an amazing legacy if you think about it.

"There's one notion that the Valley will become an urban center. I'm not convinced of that, I think the Valley will exist in both the physical world and the virtual world," said Malone, who lives in Santa Clara County's oldest home, the historical Murphy House. "I think the Bay Area will become denser in terms of population than it is now, but not much. The population will continue growing until the middle of this century, then it's supposed to start falling. So the population may be the same or a little more 100 years from now. I don't think it will be fully urbanized. I think what will happen is that it will expand out over the hills — right now you have people living in Stockton, Tracy and Sacramento who are commuting in every day.

"Historically what happens is that companies get smart and say, 'Hey we've got this body of people sitting out here; let's put a plant out there.' My sense is that the Northern San Joaquin Valley will become part of Silicon Valley — from Stockton, Tracy, all the way to Redding you'll see more and more industrial parks and they'll go all the way up to the gold country towns. We're already seeing the Sacramento and Roseville areas surrounded by Intel and HP. It's happening in Santa Rosa too. It'll spill over to the coast — Half Moon Bay, Santa Cruz... most of all it'll spill through Hollister, down to San Juan Bautista on its way toward San Luis Obispo. I don't see that as a giant megalopolis; it'll be more like Cupertino — a lot of industrial parks, housing developments, an occasional fairly tall office building. A lot of fiber optics in the ground, light rail and eight-lane freeways. I think that's the physical Valley.

"I think the Virtual Silicon Valley is going to reach around the world and it's going to interlink the pockets of technology everywhere. We are already linked to Seattle, L.A., Boston, Northern Virginia, New York and Bangalore, India. Oftentimes people in underdeveloped countries, who are in these enclaves of developed world are working for Cisco and other companies, are sometimes more Silicon Valley than some people who live in Daly City. They talk to Silicon Valley 25 times a day and they work for Cisco or Oracle and they have the company badges, get the newsletter. Their bodies may be in Bangalor or Jakarta but their heads are in Palo Alto, San Jose and Milpitas. I think that kind of virtual Silicon Valley will spread out in a network around the world. I think the reason it will be Silicon Valley is because those people will all exhibit the Silicon Valley personality: entrepreneur, tech driven, risk takers. The legacy of Silicon Valley is to establish the business paradigm of the new millennium."

John Warnock has already built up his virtual workplace network worldwide and feels that physical location is far overrated.

"If in your planning process, there's a big component about where people are located or co-location of something — it's probably wrong thinking. Because place doesn't really matter; we have development groups in India, Hamburg, England, Boston, Minnesota and here. They're trading files on a nightly basis, and where tech support is located is completely irrelevant. It's where the labor source is. This is one of the toughest places to be. It's very, very hard to attract new talent and retain the talent; it's much easier to start up an office somewhere else because electronic communication and the way the company is run is just not relevant to where you are located.

"Because place doesn't matter, I think the Valley will distribute to the rest of the country. What's important to the Valley is the mindset it really is okay to bet money on risky situations and it's a prosperous place for new ideas. And that's a mindset. If you go to the Midwest, you don't have that, you have a fundamental conservatism about risk and failure. There are counters to that — there are startups everywhere — but that's sort of the Valley's mentality getting exported. Software companies on the East Coast are different from software companies on the West Coast — they are different animals, but I think the success of the Valley is going to become contagious and spread. I've been on the boards of companies that are located elsewhere and sometimes they just don't get it. They just don't understand how it works."

CHANGING OF THE GUARD

Although virtual offices worldwide are catching on — in this place and time — Silicon Valley is still spreading in all directions to make room for new arrivals. Santa Clara Valley, once known as a suburb of San Francisco, has made a complete about-face. Silicon Valley encompasses all of the South Bay, into Gilroy, Scotts Valley and Santa Cruz, and up into the East Bay, San Francisco and north into Marin and Petaluma. It is shifting to make room for the homes and commercial space needed into the next millennium. With these swift strides in the movement of tearing down, building and spreading, people who've had roots in the area will be displaced. It's a lesson learned long ago: casualties occur in a growing economy.

Coyote, an area just south of San Jose, seems to be a beautiful place to live. An occasional crop duster may fly overhead or an automated vegetable harvester or tractor may buzz in the distance, but from the hills it's not difficult to imagine being 100 miles away from Silicon Valley instead of 25 minutes. Golden rolling hills dotted with ancient oaks plunge into lakes and dams and give a palatial sense of what the Santa Clara Valley used to be.

With the exception of the 18-wheel trucks constantly moving down Highway 101, the place seems to be suited

Photo by Sally Richards

for those wanting a quiet way of life. Valleyites driving through Coyote on their way to a weekend getaway in Monterey might sigh out loud thinking about all the expanse of land surrounding homes in Coyote, in comparison to their own zero lot townhouses. What few outsiders know is that the city is slotted for change. Cisco Systems, one of Silicon Valley's largest corporations, has its sights set on the miles of attractive land.

In the 1950s as the Valley of the Heart's Delight began the shift into Silicon Valley, farmers in Coyote began to take notice. As urban sprawl became imminent in 1958 after Coyote was annexed, farmers not only had to work within nature's blessings and disasters, they also had to come up with a way to pay the newly rated city taxes. Land belonging to families for generations was being parsed and sold acre by acre as farms shrank to accommodate farmers' budgets. Developers purchased the land, acres at a time, patiently waiting for the next lot to go up for sale. These developers, many who had made their fortunes in buying and building Santa Clara County, bided their time and leased the properties back to farmers until a day when Silicon Valley would need the space — a time they knew would eventually come.

Now that Cisco and other companies have been negotiating with the politicians and the environmentalists over the details, Coyote is about to become a part of Silicon Valley's history.

Photos by Sally Richards

LONGING FOR A SIMPLER TIME

Johne Baird is a cattle rancher in Coyote whose family owns 2,500 acres and 200 head of cattle. He lives in the California Victorian ranch home his great-grandparents built in 1916. Prior to that time, his mother's grandfather had owned a ranch in the 1840s in Gilroy.

"When I was a kid, we knew everybody in the town," says Johne Baird of how things have changed in Coyote. "It's definitely grown. You used to go to the grocery store and you knew everybody. My dad always says that people come from the land, or from a farm, and if you look back somewhere in your background your people came from the land. We've gotten away from that; I don't know how else to explain it.

Photos by Sally Richards

"There were a lot of different activities for kids; places to take animals like horse shows and county fairs," Baird said as his young son and his nephew played on a stationary wooden steer. "Now they say that you have to have activities for the kids to keep them out of trouble. Well, when I was a kid, there were all kinds of family activities… and there were chores. If you wanted to entertain yourself, you went out and dug a hole somewhere or walked out in the hills."

Baird spends a lot of his time these days lobbying for farmers' rights. Many of the people who have moved to Coyote from Silicon Valley have tried to ban the spraying of pesticides and the use of noisy farm equipment. Now, says Baird, when people move to Coyote they sign a paper that releases farmers from future lawsuits. "City people think of farming as an inconvenience; they think, 'My bell pepper comes from Safeway.' Well, they don't come from Safeway, they come from a field. Most people on ranches now get very involved in issues that face them. You can't stay home now and take care of business like you used to be able to do. One day you might have to go clean out a spring, a water system; well, you can't do that because you have to be in San Jose defending your rights because you're a rancher. You've got to get involved. There are so many issues that face us; it's more an annoyance

change in his time. As a young man, Harold worked for the state in a stockyard in San Francisco, but in 1951 moved back to the ranch with his new wife.

"Everything in the valley was all prune trees, and you wouldn't see a house except for one maybe every 10 or 20 acres," said Harold Baird of how the landscape has changed in his lifetime. "The trees gradually were all pulled up. There wasn't too much traffic. I don't think it's changed for the better. I can still hope my grandson will grow up and stay on the farm, but how long will it last? They want to keep a lot of it in open space, but how long can you do that when you've got all these people who need to live someplace? They're [people] not diminishing, there are more and more. Even the birth rate would take care of all the land today if they let people build on it, let alone all those people they let come in. I think it'll be just the same way it has been in the past — what people want than anything else. And if you don't do anything, then you get it shoved down your throat anyway. I've got a driveway that's been there since 1918 — there has never been a planning commissioner drive up that driveway and say, 'What do you have planned?' It's always been, 'This is what we have planned for you.' It's not like 30 years ago when we were the majority.

"We've been here long enough that we want to stay. I'm very lucky to have a life that I like. God makes more people every day, but he doesn't make more land. We've got to figure out where we're going to put these people, but we're also going to have to figure out how we're going to feed them. I don't have the answers. I hope my son will be a farmer, but sometimes I even think about doing something else."

Harold Baird, Johne's father, is a tall man with a serious look about him. It seems as though he's seen a lot of

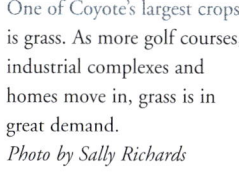

One of Coyote's largest crops is grass. As more golf courses, industrial complexes and homes move in, grass is in great demand.
Photo by Sally Richards

Photo by Sally Richards

is what they're going to get. It just depends on what their philosophy is. You talk to the board of supervisors about what their philosophy is now, it doesn't mean it's going to be that way 10 years from now when there's new people and a new way of thinking. Hell, 50 years ago you couldn't have told these people things were going to be like they are now. There's no way of predicting the future. You can have your own ideas of what should be, or what it might be, but it doesn't mean that's what it's going to be."

GROWING TRADITIONS

Row crop farmer Nicholas Jones and his brother Earle come from a long line of people who have made their careers from the soil. Nicholas takes care of the crops and Earle manages the High Rise Farm produce store on Palm Road near Santa Teresa Road in Coyote.

"My grandfather was an immigrant from Yugoslavia who came here at the turn of the century," said Nicholas of his family's heritage. "He started with cherries, apricots and prunes, which was what all his buddies were doing. He purchased land that became extremely valuable because of the development of Silicon Valley and the next generation; basically my uncles lived off of the value of the land and made tons of money in the 70s."

Nicholas and Earle Jones
Photo by Sally Richards

Deciding to get an education he may one day need to fall back upon, Nicholas attended the University of Santa Clara and studied mechanical engineering. "I worked as an engineer for eight years and didn't particularly enjoy it. I thought I'd get back to my roots. Now, I grow row crop vegetables... bell peppers, chili peppers, broccoli, lettuce, onions, cucumbers and pumpkins."

Nicholas and Earle don't raise fruit as their grandfather did. They can't even claim the land they grow on. The Joneses lease the property from a developer. Each of their 800 leased acres produces roughly 25-50 tons of produce. According to Nicholas, the produce stand sells only one 100th of a percent of what they grow.

"There are general trends in the industry of consolidation and there are fewer and fewer large bulk buyers of product as more canneries close and consolidate," says Nicholas. "It's not just canneries doing this, but freezer companies and fresh processors. They want to do business with larger vendors who can meet their demands for year-round supply and wide product line. You have to have depth and breadth; it [the industry] doesn't suit itself to family farming. You have to be operating in Mexico and here, and it takes a tremendous amount of money. Nothing goes by train anymore, everything goes by truck."

And since most of the canneries in the area pulled up stakes a while back, with Del Monte being one of the last to move its operations to the Central Valley, trucking product has become more costly as the miles from product to processor increase.

"Houses and industrial parks... it bothers me on a personal level because it's affecting me personally, but that's just the way it is," said Nicholas, a man who seems to know economy's limitations. "Farming can't compete with so-called 'higher uses.' Agriculture is a dying industry in California. There's really not a future in it. I'm trying to figure out what I'm going to do. It's not just here, it's all over, [agriculture] it's sick all over the country. There's just a lot of structural change and pressures. There's no room for the smaller mid-size grower. You can either be a hobby farmer and try to sell to restaurants and farmers markets, or you can be a giant conglomerate. I don't see anything left in between. In terms of what's going to happen out here, you can just look at the numbers. I'd say it's three to five years before this particular zone is paved. I don't know if I personally have another year left out here. I'm losing large amounts of money.

"I have no message for other farmers. It's just a very unhealthy sector of the economy right now," said Nicholas, who seems to have come to terms with destiny. "There are a lot of people moaning about how there's no strength to the marketplace. People are just hurting financially and a lot of them are going to get squeezed out. That's the stage of the cycle we're in. I have no great message; I'm looking for messages from other people. At the moment, at least here, I don't see a future for family farming unless you happen to inherit a large corporate-scale farm," said Nicholas as he bid farewell and climbed back into his truck to manage the day's harvest.

Earle Jones feels that farming in this country will go on, just elsewhere. He longs for a time when farming was younger and farmers didn't have to compete with other countries with unfair advantages.

"How many places in the U.S. or in the world are really suited to grow cherries and citrus? The U.S. is one of the few places you can grow some of these things. If you want to look at what we're losing, it's a lot of the more specialized climates that don't exist elsewhere. Climatewise, Mexico has some similar areas, and as things can't be grown here they'll be imported from elsewhere; it's just supply and demand. As this area is paved over to make houses, fruits and vegetables are still in demand. If it can't be grown here, it'll come from outside the country where perhaps the laws are more lax labor-wise.

"I find it interesting that we have regulations that our government has seen fit to put in place regarding chemicals, labor laws and sanitation, we have to produce and sell under those rules, whereas growers in other countries don't have near the regulations we do.

"I started farming when I was 5, and Nicholas started when he was 3 — that's how old we were when they stuck us in a tree and we started picking cherries in Sunnyvale. I was brought up thinking farming was a way of life, it wasn't a job. I just did it. I never thought about finding anything else. I hardly do any farming now, it's mostly handling all the retail here on the farm. Basically, people just want something that tastes good. Most farming has all gone to vast bulk production; it's not grown to eat, it's grown to ship.

Dot-com companies are flooding the Valley with billboards, even next to ranches in Coyote.
Photo by Sally Richards

Photo by Sally Richards

"Quite a few people don't feel it meets their needs as far as taste and nutrition. It's not just taste, if you pick it early it's not going to have everything that should be in there.

"Most people who come in [to the store] don't like to think about it [urban sprawl]; everyone is too busy to sit back and think about it. But what might some alternatives be to paving things over? This area — the Santa Clara Valley — has such diverse microclimates and there are crops you can grow here that don't do well in other places. The most interesting places to grow things are all under pressure from urban development. How many zillion acres across middle America can you grow wheat or grain crops... corn, cotton things like that?

"I think farming will go on somewhere," said Earle Jones of the future options for Coyote farmers. "Coyote will be paved over sometime in the not-so-distant future. It's a shame that we haven't found a way for some farming and open space to coexist with the total suburban sprawl in Santa Clara County. Part of it is that there's just so much money to be made building houses and businesses and nobody has really come up with a solution. In Marin County they've bought out the development rights of farms, so the farms can be bought and sold, but only as farmland. Maybe there needs to be some thought about how we want to live. If everyone is happy with having every inch of the Valley floor paved over and gridlocked, that's how it'll be. But, I'll have to go elsewhere because I'm fond of open space and not being boxed in. There are a lot of people who feel that way. It's always sad to see the ground paved over because the ground is productive.

Stanley Moniz
Photo by Sally Richards

"Years ago people didn't work all the time, you had time for a barbecue and it was a different way of life. Now everyone just stays home and can't do anything on the weekend because they have to take care of the chores they couldn't do during the week. I don't think of the changes — not just to do with farming or lack of farming in the area — as progress. I wouldn't say in a lot of areas we've progressed in the sense that we've moved forward or to a better place; it's just change, some of it for the worst. That's just my opinion," said Earle Jones, as he smiled, shrugged his shoulders, replaced his hat and went back into the store to get on with the day's business.

VINES GROW DEEP

"In 1960 it was all vineyards," said Stanley Moniz, a Zinfandel grape grower near Coyote, of the property surrounding his 10 acres. "We took vines out in order to dig a well and build a house, and this is where we live today." His vineyard was originally planted in 1917 and has won numerous awards during its lifetime.

Moniz, born in Morgan Hill, graduated from high school in the 1930s. He began his career implementing farm and dairy hardware in the San Joaquin Valley and then worked as a plumber until he retired in the 1980s. Now, with his son, Moniz raises more than eight tons of

grapes a year for wine hobbyists who buy a quarter to a half-ton of grapes at a time.

"I don't plan to move, this is our home," said Moniz, pulling down on the bill of his baseball hat as if in defiance of anyone trying to move him. "Other than the freeway that came through here eight years ago, it's ideal. I used to see miles of prune orchards from here. I'd like to pass this property down to my son, my granddaughter, grandson and my great-granddaughter. But we've had an experience

that makes me wonder about this. We had a member of the family who had property in Campbell die. We kept the property, but the IRS took an absorbent amount of money on death taxes, so it depends on what they appraise this land at. If they appraise it at a million an acre, we have 10 acres... well, I don't know what will happen.

"When we first moved here, we didn't know anything about the freeway coming in. It had been planned, but we didn't know that; it probably wouldn't have stopped us from buying." Moniz pointed at the back of his property, "This place used to be different, now you can see a Target and a Mervyn's sign. The property in back of the vineyard is now up for sale... for development. For us, I don't think we'll be affected too badly. The big thing is going to be the traffic. It's difficult to get on the Monterey Road. The freeway — when they bottle in here in the morning, it slows people down. The Coyote Creek runs along the base of the hill, that's Anderson Dam up there if you could see beyond that big building. When we first moved here I could go down to the creek at night and catch two, maybe three native trout; today you can't do

that. They drained the dam to repair it and remove trees at the bottom. They didn't have a source of water to maintain the trout life.

"At one time, there was a dairy behind me," pointed Moniz to property where buildings now stand. "All this was a pasture for the cows. Around February and March mushrooms used to grow here. The original owner passed away and his son sold it to a developer. I miss the peace and quiet we used to have here."

The Valley's Future

As Silicon Valley grows, so does its work force. For good or bad, land prices are rising and people make more concessions to live in the heart of the world's technology center. Many would do absolutely anything to be in the proximity of this cutting-edge technology and billions of dollars in venture capital. Anything.

There seems to be no end in sight to the extremes entrepreneurs will take to be part of what is happening. Nightly news that used to carry the ordinary crimes and disasters, now headlines with IPOs and mergers. Jeff Bezos, the CEO of Amazon.com — a high-profile company that has yet to see black ink — was on the cover of *Time* as its "1999 Person of the Year." The president of a company called Docusearch.com, an online private investigation company that investigates people who entrepreneurs and venture capitalists might think of partnering with, was on the cover of *Forbes*. At Denny's and Starbucks people are talking about business plans and stock options.

In today's economy, advertising directly targeted to Silicon Valley's audience is everywhere.
Photo by Sally Richards

Moniz Vineyard
Photo by Sally Richards

Photo by Sally Richards

As in any industrial revolution, there are people now emerging as the leaders of the Valley's new economy. This new currency is based upon bandwidth, speed, branding and integration. Millions have flocked to the area to seek their fortune in this new-world economy where the average age of a millionaire is 20-something and where serial entrepreneurs start companies as easily as a bee going from flower to flower to pollinate.

Many of yesterday's entrepreneurs have become angels and venture capitalists and are now richer on this last rush of dot-com businesses than they were in their entire careers as inventors and innovators. The saying "Old CEOs and entrepreneurs never die, they just become venture capitalists," applies perfectly. According to *Business Week Magazine,* in 1996 one company went public every five days and 62 Internet millionaires were added to the gilded roster every 24 hours — and it has shown no signs of slowing down. Mercedes Benzes, Jaguars and expensive SUVs are more common than not in the slow traffic snaking its way throughout the Bay Area.

Many of the new millennium's dot-com companies might take a lesson from the words of wisdom left from other eras as a reminder that this whole business is probably going to be cyclical and prone to ups downs and utter saturation.

"I think probably the opportunities are not as good as they were 10 years ago for new starts," said Russell Varian about the business of startups in a 1958 taped interview housed in Stanford's archives. "Though I think there are quite a few good opportunities for new starts. I think if anyone is considering starting up on his own with small capital, he has to look very carefully at the field and the products that he intends to exploit. He has to have something that has a potential for growth, but not too rapid of growth. If it expands very rapidly, well one of the larger companies will undoubtedly take away the business. There's a definite place for the small company in new fields. Any one product tends to saturate at a limit that is too low for the big companies to be interested in. The small company can out compete the big company in these fields because it's more flexible, the management is much closer to the products and the development. Usually, the management is experienced engineering-wise and not as encumbered by the red tape a large company gets into. I think there still is a field for the small company in which a large company can't compete."

THE TIME IS HERE AND NOW

As the farms and ranches become casualties of the new millennium, high tech is still trying to catch up with itself, to find its footing in this world that moves at the speed of light. This is a time of tremendous growth, and part of the success of any company is establishing relationships with clients and customers — it's also partnering with the surrounding community. Corporate philanthropy brings people opportunities they may not have had otherwise. This new era of giving has at its helm one woman in particular who believes that part of giving back is making sure the next generation of women has better opportunities, and viable places to work and live.

"The very interesting question is whether Silicon Valley will be a Detroit 100 years from now," said Carol Bartz, CEO of Autodesk. "Will we continue to be the innovators? How long can a place stay on top? Which companies that were around at the turn of the [20th] century are still viable? GE? There are a few. Which parts of the country are still viable? If you look back 100 years ago in the mill towns of the East Coast and now, 100 years later, it's tipped over here to the Left Coast. And so maybe you have to think it'll be somewhere else, maybe another country. We have to prepare on how to stop that. The premise of that is back to a much narrower view, the whole educational process, the whole process by which people believe they're valued human beings. You have to have a group of folks who believe that they can fight for anything — that's what propelled this country the last 100 years. Those people felt as though if they worked hard they could get it — and you could make a difference. I think we've lost some of that, because we lose it right off in the education system.

Photo by Sally Richards

"I spend a lot of time saying, 'Laney, life isn't fair. Deal with it.' I believe that one of the ways to deal with it is to be prepared," Bartz says of her 11-year-old daughter. "My consistent message to young girls is to be prepared to have a choice to do anything. My dream is to have every girl pass college freshman calculus, because it gives a woman the basic fundamentals to do anything she

wants — to go into engineering, to go into the sciences. Who cares if she then goes into dance? She's a smarter artist. If she can't do that by then, 60 percent of careers are just cut out of her options. That's your jumping off place for most degrees and I think for confidence. My message to the parents and their girls is, 'Get yourself to that position and then you'll have a choice.' That doesn't mean that we don't have to do a lot of work to make sure there's not a lot of bad external influences of people saying, 'You don't need to do that.' You have to work on

Carol Bartz is a mentor to girls all over the world.
Autodesk

Photo by Sally Richards

all that stuff. I'm not saying we have to change society, we just have to know what our responsibilities are.

"I think women in general have to be ready to try things. Too often they believe they have limitations, or they go for perfection instead of getting out of their comfort zone and not being afraid to fail and just going for it. It's easy to say 'Hey I do this well,' but 'I'm not sure that I'd do that well.' You have to go for it. Hey, if it doesn't work, you do something else. You have to have the attitude, 'Why not?' That's what I keep saying to my daughter, 'Hey Laney, try it, just go for it.' I can't tell you how many times she says 'Hey, that was easy.' Finally, it becomes a very comfortable thing to do.

"I worry about what will my daughter see, not what I will see; I'm already 50 years into this problem. You have to assume that she'll have a life span of 120 years, so she could actually see another century change. I worry if we don't get the have and have-not problem out of the way, it's going to be uncomfortable for everybody.

"It's very clear that all of us in technology are worried about future employees," Bartz says about a company's responsibility to help the community grow with the wealth of the companies in it. Autodesk has implemented several outreach programs where employees receive comp time for working in the community, even if it's only in their kids' schools. The company also has a college scholarship program and several mentoring programs geared at college women and high school girls, including the award-winning Web site *Design Your Future*. The site is managed by young women in high school who intern at Autodesk and have designed everything on the site (code, graphics and content) with the goal of educating other girls about careers in high tech and discussing common concerns and interests.

"There are many companies who have to train basic in math and English skills," says Bartz on the level of skills some schools are graduating students with. "I would like to save that for the job of the schools. And if we all did our jobs to the extent where we could add some realism — either by encouraging girls to really study hard, or job sharing and shadowing and understanding why it is important [to learn math and science] — we should be able to help. I think a lot of companies are coming to those kinds of conclusions. If you can get someone interested now, at a young, impressionable age — we have insured our future. This week, I received an e-mail

from a girl in Hungary who found our homepage and saying, 'I wish I could do that....'"

The letter sent to Bartz was from a young girl who had visited the DYF homepage. Although the letter is written in broken English, the message comes through loud and clear. The world is becoming a much smaller place and Carol Bartz continues to change the world one girl at a time.

> ### Dear Carol Bartz:
>
> My name is Edina Varga. I'm not a US citizen. I have born and live in Hungary, in a little country in the heart of Europe, in that country, which has already given a lot of talented scientists to the world, like Janos Neumann, Ede Teller, and so on. Unfortunately, I cannot mention any women, yet :)
>
> I decided to write, because I'm extremely interested in math, science and technology. And I was very happy, when I have found this homepage. I know, that my possibilities are quite limited, because of the great distance. And I'm not so lucky, that I could participate in a job shadow program, even if I find it very fantastic. But if we could change some thoughts, I would be extremely grateful. I'm still at high school, I'm going to make my graduate in this year. I have already mentioned, I love maths. It was always my favorite and fortunately I also have a little bit talent in this field. Why so many girls don't like maths? In my opinion there are very few good maths teachers. I don't say, that they are not qualified enough. The problem is somewhere else: they don't know how to show the beautiness and greatness of maths. On the other hand, there are too many prejudices. I mean, that most of the girls say, that oh., I don't have any talent to maths, instead of trying to understand it, and practicing more. I think that we should also learn more about the history of mathematics. In order to understand, that this science is not only about numbers, but has more thousand years old history, and tradition. Let me tell some word about myself. I'm still at high school. As far as my plans for the future are connected, I'd like to be an electrical, or mechanical, or computer engineer, or mathematician. For my decision I need more information about these jobs. My biggest dream in my life, is to became as good in my future 'mission' as I can. And I'm ready to do everything for it. The world is like an enormous tree, and we are the leaves on it. The life of a leaf is ephemeral. They born in spring, and die in autumn, but the tree's life goes on, with new leaves and new lives. My aim is to became a very clever leaf, in order to give to the humanity as much as possible. I'd really like to know, how could I get involved in your work.
>
> Hope to hear about you soon:
> Edina Varga

The Dot-Com Generation

Money is running thick in Silicon Valley. Everywhere one goes there's a deal in the making. Much of the deal making is done at restaurants like Buck's (where Netscape was founded) in Woodside or Café Barone (where Marimba was founded) in Menlo Park, and other places where entrepreneurs can gain audience with a perspective

In Silicon Valley, dot-com advertising is everywhere one looks. Cost is no object when branding a new Web site address.
Photo by Sally Richards

Drivers traveling north on Highway 101 on the Peninsula have seen this billboard for nearly a year. Some companies pay more than $5,000 a day for the 101 placement. This *Forbes* ad is aimed at the Silicon Valley dream: the dot-com professional retiring at 20-something.
Forbes

Craig Newmark with his craigslist.org team
Photo by Sally Richards

angel or venture capitalist. All you need is a killer business plan and an executive team, and you're in like Flynn.

"The stock evaluations are going to crash, but the Internet certainly isn't going to crash," said Warnock of the trends he sees with the new economy. "It's in vogue today to not have earnings and to go for market growth and penetration, but at some time you have to have a long-term business model that's profitable. So if you don't have a clear vision on how the product is going to relate to the market and its customers and what the value is you're delivering — and if its sustainable — you're in for a very hard time," says Warnock who heads Adobe's venture capital arm. The fund that has fed the driven entrepreneurs of the new millennium to the tune of $90 million over the last three years has received a $260-million return on its original outlay.

Many of the dot-com successes are planned and implemented with great care and panache, but one stroke of Internet genius happened nearly by accident. Unlike many dot-com companies begging for capital, craigslist.org has turned down a lot of capital and buyout offers from large companies. Why? Well, Craigslist.org is one of those rare dot-coms making revenue.

"I grew up with a pocket protector and thick glasses taped together and I grew up excluded," said Founder Craig Newmark. "Now I'd like to see a network of craigslist sites across the world including everyone and giving everyone a voice to do what works for them in their particular culture." Craigslist has become the site for the trendy dot-com and established tech companies to advertise job listings. Newmark, a programmer by trade, started the company to help people find jobs and housing in the Bay Area. The company garners many millions of hits — and page hits are the currency of the Valley — a month. It has also become *the* place for creative tech writers, code jockeys, dot-com marcom experts and the like to hook up and discuss their issues, form alliances, but most of all make friends.

Craigslist sponsors parties where the hippest, pierced 20-somethings and socially handicapped geeks finally get an opportunity to meet and put a face to the e-mail. These events have set off partnerships and romances, but most of all they solidify the virtual community into physical form; something that no other site has managed to do as successfully. The branding of craigslist — something that companies pay millions for and may not ever succeed — was practically accidental. It was one guy, Craig, in a Southwest-decorated flat in San Francisco who followed his dream of giving everyone a place to fit in. Craig has become so well known that people rarely know his last name — his villagers know him as Craig. It's the kind of success found in Hollywood with other single-named people such as Cher and Sting. Craig was recently named one of the Top 10 bachelors in Silicon Valley — he's come a long way from his lonely high school days.

The company never accepted a cent of venture capital but is growing substantially — even more so than the sites that have millions in VC. "There potentially is a place for everyone who provides some kind of service, people who provide some real human intelligence. In many cases, if they're not providing a real human service, people are going to take their clicks elsewhere."

Not every entrepreneur in the Valley starts a successful Internet site from happenstance or is a "newbie" fresh out of college with an MBA. William Lohse is a veteran of the ages, steeped with the wisdom of what has come before and a fairly vivid idea of what will follow. He's one of those serial entrepreneurs whose successes keep him coming back for more. With a bachelor's degree in philosophy, he began selling vitamins door-to-door in Sausalito in the mid-70s. It was at a front door cold call where he met someone who admired the tenacity of the 23-year-old vitamin salesman and offered him a job in high-tech sales. He rose quickly to the top as the president of the European arm of the company. After leaving that company, he became the first director of sales for Wordstar, the killer app of its time.

There were other companies and achievements, but Lohse would go on to join Ziff-Davis publishing as the publisher of *PC Magazine,* at the time known as the Bible of the computer industry. Over the years he learned a great deal from his mentor, Bill Ziff, and in true Lohse-style, he rose to the top as president of Ziff-Davis. After leaving ZD, he had a successful stint as an angel, but Lohse wanted to jump into the fray of the dot-com excitement coming down the pike. So in 1998 Lohse started SmartAge.com, an innovative company to help small businesses build important b2b (business to business) relationships. After raising more than $50 million in capital, it wasn't long before he was going head to head with one of Bill Gates' companies for the small-business market share. And Lohse seems to be winning; SmartAge is the largest company in its category with more than a million members and 3,000 new members a day.

"Bill Ziff taught me to be early and to be lucky. One of the clearest ways to be lucky is to figure out a category that's going to win and get in early," said Lohse of his timing. Lohse has long been looked to as a visionary, a man who tempers science with philosophy. His ideas may be a bit futuristic, but he sure saw this era coming. "The future is anywhere, anytime as much bandwidth as you want. We're going to push sand as far as we can and then there will be a biological effort using cells as gates — the wet chip. We'll be able to program anything we want — IQ, useful looks, longevity, but this could be the worst kind of bifurcation between the wealthy and everyone else. It could be someone's plan that we become some sort of soulless wet robots to develop value. Time is so real, but in the next revolution we'll realize we can manage and manipulate it. We can't even imagine what the information age is going to be — what is my grandchild going to do with that?"

One of the venture capitalists who saw promise in SmartAge.com was Christine Comaford, one of the highest-profile VCs in the Valley, who is barraged with

William Lohse, CEO, SmartAge, is always building something new. *Photo by Sally Richards*

stacks of business proposals seeking funding every week. CEO of Artemis (the goddess of the hunt) Ventures, Comaford's mantra is "money is cheap, time is expensive." The company's portfolio includes winners such as AGAiN Technologies, Eletter, Clairvoyant and toolwire. In today's e-conomy how can you tell if the next big thing is going to be b2b, b2c (business to consumer) or something else all together? How do you know who to invest in when you're responsible for doling out millions? Comaford came into the business the same way as Lohse, up through the ranks (as a programmer), through successful entrepreneurial startups and finally as an investor. Through this hands-on life training, she seems to have emerged unscathed and empowered.

"I make decisions with my heart, mind and gut — each gets a point. We only like to get involved with painkillers, not vitamins. It can't be optional, the b2b

Venture capitalist
Christine Comaford
Photo by Sally Richards

Photo by Sally Richards

market is this — you need to offer a good solution for the pain a business is suffering, so much so that if you offer them the solution, then threaten to take the IV away, they say 'no way.' Once your company gets inside of the company, they find additional pain points they may have not even been aware of when they were on the outside. You also have to find out who these humans are you're getting involved with. Can you see them going through heaven and hell and will they be fun in both venues? Even in good times there are problems; growing too fast can kill growth potential. Even massive success can create massive problems.

"What we're going to see is consolidation — we don't need 50 million pet and a thousand grocery b2c sites," said Comaford of the oversaturation in the market. "What we're finding is that people like to buy stuff all at once, not go to sites like socks.com, deoderant.com to buy just one thing. There'll all kinds of companies with profit in 2000, but new crashes and consolidations... powerful players will emerge. The government is entering the game and trying to intervene, but these big, powerful players are going to say, 'get out of the way, this is my backbone and you can't slow me down.'

"The hardest question about the future is if we'll all be living longer in some kind of plastic body. We'll probably see interplanet 'beam me up Scotty' kind of travel in 100 years... wild stuff. But in the next five to 10 years, the markets that are percolating now — Denver,

Boulder, Austin, Boston, LA, Atlanta — they won't be just baby Silicon Valleys, they'll enter the era as powerful players. Silicon Valley can't be the center of the universe forever."

On the flip side of venture capitalists is a popular breed of investors called angels. It used to be that angels were private investors who had no way to really do thorough due diligence about the technologies involved in investing in disruptive technologies. One group who figured this out is The Angels' Forum, a group of people handpicked by Managing Partner Carol Sands, who have expertise in various disciplines (manufacturing, science, technology, marketing, etc.). Once she filled the chairs at the table, she opened the doors for the submission of online business plans. Very few get through the door physically to make a pitch to the group, but those who do have been investigated thoroughly for viability in their marketplace.

"A good angel investor, by my definition, is local," Sands said of her group's investment strategy. Many of her angels spend physical time as mentors at the companies they invest in. "I also believe Silicon Valley is one of the best places in the world for a woman to be working. Although the Valley is not totally free of gender biases, Silicon Valley is miles ahead of other communities. I see and meet more entrepreneurial woman leadership roles today than I did two years ago.

"I believe the rate of adoption of technology will, in fact, increase," Sands said of future trends. "The concept of living in an environment of constant change has become accepted by our society. The long-term winners of the global technology race will be those companies who figure out how to make it easy for new users to adopt and use their technology."

What does Sands see for the next 100 years? "Just as Florence and the Renaissance are thought of together, I think historians will look at Silicon Valley and this time as the beginning of the Imaginative Technology Age. We will continue to create, adopt and normalize technology for a long time. The pilgrimages to Silicon Valley will continue, and each pilgrim will take their interpretation of our economic and social values back to their homeland and the impact will begin — just as it did back in the Renaissance.

"I have great hope for the future. I believe I will have good friends that I see and socialize with regularly from all over the world thanks to travel and time technology advances," Sands said with enthusiasm about the world she will help to build, one company at a time. "Space travel will be commonplace. While global laws will be focused on human rights and environmental concerns, uniqueness will be prized. The biggest shift, beginning toward the end of this next century, will be the impact of our significant advances in health management. Death, aging and religious values will become the issue of the next few centuries because the suffering that traditional economic class separations caused by the lack of food, shelter, clothing and education will no longer exist. In short, I believe that technology will cause humankind to evolve."

Another woman in Silicon Valley who is making positive changes is Dixie Garr, VP of Customer Success Engineering at Cisco Systems, a woman some say will one day move into John Chambers' office. "I think there are

Dixie Garr
Photo by Sally Richards

so many things we can't even imagine," said Garr of the future she sees for the Valley. "I think in the not very distant future, the epitome of the major change for women is going to be when a woman comes to head a major company in Silicon Valley — and there will not be all the publicity about 'so she's a woman and what does her husband do?' It will just be common knowledge and

Photo by Sally Richards

accepted as a natural part of life. That will lead to all of the other roles changing as our natural leadership and participation on an equal playing field happens. This whole trend in technology puts us on an equal playing field. We have the connections and the knowledge it takes and we will be there as equals in — I predict — the next 15 years.

"My observation is that parts of this country are ahead of Silicon Valley in terms of their awareness of women and diversity and inclusion," Garr says of what much of Silicon Valley still needs to work on. "I think the Valley is moving so fast that we don't often have time to think about it. And it is something that requires thought. To pick somebody who is not like you [hiring], you really have to concentrate on it not to do it, it's an unnatural act right now. In other places where companies have matured and they're up those learning cycles, they are now starting to raise their heads above water and say, 'how do I prepare for the future?' Silicon Valley is full of start-ups that don't have that leisure. I think that some other parts of the country are significantly above us, but I don't think that's going to last for long. The reason is Silicon Valley may not be the first one to notice but when they realize it has an impact on business, they're the first ones to lead. I think they're going to get there first, they're just not first right now."

VISIONS FROM THE FUTURE

One of this era's futuristic dreamers is Kim Polese, CEO of Marimba. Polese left Sun after pushing Java into a successful market position and started a company of her own. Raised in Berkeley, Polese vividly remembers many trips to Lawrence Hall of Science where she interacted with a computer called Eliza that ran one of the first significant artificial intelligence programs, an online virtual psychiatrist. It must have seemed like Arthur C. Clarke's creation HAL9000 from his novel *2001: A Space Odyssey*. The young, wide-eyed Polese was sold, hook, line and sinker, on technology.

"I feel like its just the beginning. I don't think we've even begun to see the kind of accelerated growth that the tech industry will experience in the next century," said Polese of the future envisions. "When you think of nano technology alone, we're just starting to have cells think and reason — cells within the human body will be able to do intelligent things. I think we're at the very, very beginning and I think the pace will continue to grow. I don't see it ever slowing down in my lifetime. I'm looking forward to these new technologies developing over the next 50 or 75 years or however long I live; with life expectancy increasing, it could be 100 or more by then. You can't even conceive what will be part of our daily lives. That's what so exciting about being in Silicon Valley right now, you can feel it happening.

"The marriage of biology with silicon — we can't even conceive the kinds of inventions that will surround that, we're just at the birth of it. In college I remember becoming fascinated by this idea of neural nets and the marriage of the physical world and the digital world with the biological. It was a dream back then, but now it's starting to be real and the first companies are starting to form. It's a whole new universe — a whole new world. I majored in biophysics in college for exactly this reason. I found the intersections of different scientific disciplines to be so interesting — MRI (magnetic resonance imaging) is an example of bio-physical phenomena. The application of physical sciences to a biological phenomenon — you could see the direct impact of the technology on the human body.

excellent vantage point to see the shifts coming before the ground even starts to tremble. "I think 100 years from now Silicon Valley will certainly play an important role but not like in the extreme role it plays today; the gathering place where everyone has to come here to meet the people who are building the companies and making things happen. That's why so many people are flying out here — it's almost like a carpetbagger phenomenon. Along with all of this innovation, there is this frenzy to connect up with incredible people building these new empires and to plug into all the wealth being created here.

Photos by Sally Richards

"I hope Silicon Valley will retain its roots in innovation and creativity — and the purity of what made the birth of this technology industry so great. Creating groundbreaking technologies and changing the world was the pursuit, it wasn't about cashing out by the time you're 25, or buying a Ferrari to add to your collection, it wasn't like that. It's started to become that way; I saw it starting to

"Silicon Valley is an exciting place to be right now, but there's also cause for concern. There's an opportunity to steer this thing in the right direction and there's an opportunity to be blind and to let greed, power and ambition take over everything. I hope we will have solved that in 100 years, and the Valley will continue to be a beautiful place where innovation and creativity flourish. People will look at this as the place where it all started — a place hundreds of years from now — if not thousands of years from now — where they will want to trek and see the spot on the ground where Hewlett and Packard worked in their garage. I think it will take on that kind of historical magnitude, and people like John Chambers will be seen as the great icons who created an entire industry. We are just at the beginning of something so huge that we can't even begin to imagine its proportions, and I think 100 years from now it will be apparent."

Polese, well connected with the movers and shakers as well as the politics and politicians of the Valley, is at an

happen a few years ago — the focus shifting increasingly to money. I worry about that. The Valley is at risk of becoming known as a place of greed and consumption, a place where the ultra wealthy live side by side with people who can barely rub two nickels together. It's those kinds of things I worry about. Where all those things that made this place — like diversity, tolerance, creativity and innovation start to disappear."

What does Polese have to say about the physical layout of Silicon Valley 100 years from now? "Cars won't exist anymore, we'll zap ourselves around ala *Star Trek*, there will be changes on that proportion. I hope there's still open space — the mountains, the whole coastline — it's such a gorgeous area, and I'm fearful for the changes that are coming. When I drive down 280, I just marvel over it every day. I've lived here my entire life; I'm still constantly struck by the beauty of this place. So, I think there's a lot of work that needs to be done. Part of me hopes that 100 years in the future we will have torn down a few of these office parks and replaced them with apricot groves again. This is such a beautiful fertile place, it's so sad thinking about how much delicious fruit we could be harvesting."

Making the Parking Lots Flow

As people continue to funnel into the Bay Area, brilliant minds everywhere are putting their efforts into finding viable solutions to keep the morning and afternoon parking lots — otherwise known as freeways — flowing. One of the most brilliant engineers on the subject of transportation is Stanley Hiller Jr., aviation industrialist and dreamer. Hiller, one of the original innovators of the helicopter and founder of the Hiller Aviation Museum in San Carlos, feels strongly about solving the transportation issues in the Valley.

If anyone can solve this commuter problem, it could be Hiller, inventor of many flying vehicles including a futuristic vertical one-man flying machine called the Hiller Flying Platform designed in 1955. Hiller believes that education and a think-tank atmosphere is the solution, so he began an accredited educational institution where revolutionaries in technology and business give extension courses for credit. Hiller's university is building partnerships with Lawrence Livermoore Labs and other organizations to work on the issues of transportation, both for people and goods, and ways to generate energy and store it. The $25 million facility located at the San Carlos Airport houses both the university and the museum loaded with aviation relics that Hiller scoured the earth for 30 years to find.

"We're looking forward to the next 100 years of helping to design an infrastructure where ground, air, living and transportation work together. The changes that will come in aviation products are germane to Silicon Valley; sensors, self-diagnostic equipment and others. The electronics industry will shift into the next gear because of the way we live and transport. This is why we founded this educational institution, to find solutions that will have an effect globally."

Looking to the Sky for Answers

Everyone is looking for the answers to questions about the issues we are facing in the Valley. Wouldn't it be nice if a message from space, perhaps from a civilization where all these problems were solved a millennia ago,

Photo by Sally Richards

(Far left)
Photo by Sally Richards

Visionary Jill Tarter
SETI

Stanley Hiller Jr.
Photo by Sally Richards

would send us the answers over our own radio waves? Perhaps it's optimistic to believe it would be that simple, just in case, maybe we should keep listening. One woman who has dedicated her life to listening for signs of life in space is Jill Tarter. She knows if you listen, they will call. And that's what she does. She and her entire team monitor signals from space, waiting for someone, somewhere to acknowledge their signals. One of the most popular screensavers, available free of cost at Seti.org, allows ordinary people using computers to also listen in to the signals SETI receives.

Tarter, director of Project Phoenix of the SETI Institute, was thrown into the spotlight when the movie *Contact* was released. Carl Sagan, who wrote the book the movie was based on, knew Tarter and based the main character loosely upon her. Another reality-inspired character in the movie is the blind researcher, Kent Clark. The Project Manager of Project Phoenix and the leader of its signal detection team is a blind Ph.D. physicist named Kent Cullers. According to Cullers, "An early version of the screenplay included a small part for which I was judged competent enough to play myself. However, as the part expanded, it required the skills of a real actor." As of late, Hollywood has found plenty of grist for its monolithic billion-dollar movie mill here in Silicon Valley.

"We are such a young technology in such an old galaxy that I think there have been many encounters between different technological species — long ago and far away," said Tarter about what keeps her searching for extraterrestrial life. "We are just emerging technologically, nobody more primitive than us can communicate across interstellar distances. That's the main reason I have some confidence that any signal that is detected and has any information content will be decipherable and will not contain information that is likely to destabilize or derail our own maturation.

"They've all done this before and have probably gotten it right by now. It could be that the surest way to get from being a young technology to being an old one is to invest reasonable resources in detecting your cosmic neighbors, rather than blindly stumbling around trying to figure out how to survive your own technological infancy. It's not that I expect any answers or technological salvation. Just the fact of detecting an advanced technology is proof that it is possible to survive; that there is a way. I don't know about you, but the problems I work on most tenaciously are those for which I am reasonably sure a solution exists."

THE GATEKEEPERS

There are many inroads to Silicon Valley; some come with an invitation, others with a gatekeeper. One of the most well-traveled entrances — and one with stunning views of the majestic redwood covered ridges separating the Peninsula and the Coast — comes with a man named Jerry Morissette. He has scored millions — many, many millions of physical hits, in Valley speak — each year. He lives at a Highway 280 rest stop and hangs in the background, waiting to see when he can lend a hand. "It started a long time ago," said Morissette. "I came here to train handicapped kids to learn janitorial and gardening work [through a program] and the Cal Trans supervisor let me have a small garden area. People liked it so well that it grew and grew."

"CalTrans decided it was going to cost them a fortune if they made all the rest sites like this, so they tried getting rid of me," Morissette said of the politics of living at a site owned by a government agency. "Then 250,000 people just in San Jose signed petitions and it was CalTrans' biggest nightmare. Gary Richards from the *Mercury News* walked by and saw me and the dogs in the storeroom and says, 'hmmm, let's write an article.'" Through Richards' campaign of Morissette's plight, other worldwide media picked up the story. He was featured in *Modern Maturity* and CNN has picked up his story several times, and he is also the subject of documentaries and other press attention. CalTrans ended up purchasing a trailer for him to live in.

The classic ambulance that once was parked there is gone, but for years people were curious about why it sat there overlooking the traffic as if in the ready for a highway accident. No lurker waiting for a collision, it was just Morissette's mode of transportation, and sometimes his home. There, between the 92 junction and Black Mountain Road exit, is the statue of Father Serra pointing toward the Pacific Ocean.

"I like the statue as a statue, but I can't go along with his lifestyle. He tortured Indians in the name of Catholicism, he was a bad spot on their record," said Morissette. "He was supposed to be facing toward the San Francisco mission, but if they pointed him that way, his rear end would be facing traffic, so they turned him around, so I have this story that he's pointing to an old Indian trail that leads to the mission."

Over the years Morissette has received thousands of letters from people who've enjoyed their visit at the vista point. Many people haven't had a choice other than to stop. Some have just run out of money while crossing the country and pulled in with their tanks on empty. Morissette hooks up a lot of people down on their luck with state and local agencies to help them get back on

Photo by Sally Richards

their feet. He's intervened in nine suicides and put out nearly 300 car fires. Even with all the tragedy that pays a visit, Morissette manages to find the humor in most situations.

"One day, I heard a man on the pay phone yelling he was going to kill himself, so I called it into 9-1-1. The man left before the police came though, but all cars in the area came in, even one from East Palo Alto! The dispatcher misheard my call, and had put it out on radios that 'caretaker was going to kill himself.' When the police arrived, they asked me if I was okay, and asked me why I was depressed. I had to explain the story to nine different officers."

Morissette, once an alcoholic and sometimes lacking a place to stay, is a jack-of-all-trades. He and his three dogs and parrot make sure that people respect the privacy of others. A group of regulars stop by at various times during the day and well into the evening to sit and talk with Morissette. He listens to see if there is some way to help someone having a tough time. Sometimes he helps and he's not even aware of the gesture. "We had some European people here crying their eyes out," Morissette said of some of his recent visitors. "Apparently, I had planted a whole patch of flowers that grew in their country, in their home town. They'd been on tour for four months and when they came here, I guess it just brought home some feelings.

Photo by Sally Richards

Now Morissette, a regular on "Bay Area Backroads," has more than a dozen gardens throughout the hills in back of the rest stop and near the statue. Each one of them is a peaceful place for meditation. The physical and virtual worlds of the Father Serra rest area collide as Morissette's site comes up on the computer screen in his trailer, a home wired to reach the masses located next to the gardens. Morissette gets a ton of e-mail and fan mail from people from all points of the world he has met and become friends with over the years. The virtual Father Serra's Garden is where clickers can view pictures of all the gardens, read letters sent to Morissette and find out about the awards and commendations he and his area have received, including the coveted Blue Star Memorial.

"I know millions from being up here," said Morissette. "I just don't know a lot of them personally. It's sort of like being a priest; I don't mark anything down, there's no evidence or anything else. They come up here and take care of their problems. People feel they can come up here and talk to me and it's not going to be repeated. Basically, they come here to get things off their chest because if they do it at work they'll be fired or whatever;

they can't spill it at home because it'll get the wife upset, so they spill it on me. You get everything here from engineers to ex-nuns, from the very poor to the very wealthy. We had the president of Bulgaria stop here; he was a lot of fun. Good old Bill and Hillary were here and I was looking right down on them. They came up here in an entourage and had to slow down because they were clearing Black Mountain, so they pulled over here. And

Jerry Morissette, a modern hero to many
Photo by Sally Richards

there they were with the windows down, Bill and Hillary reading their notes. I wasn't anymore than three feet away from them, I could have tossed a rotten egg in there... their security wasn't great. A lot of rap people and country singers also stop by.

"People feel like they need a release, that's why a lot of people come here. There's always a cup of coffee here; if they need someone to talk to there's somebody here, if they feel like the world's come to an end and they're ready to reach for the flushing toilet, I can tell them how bad it really can get. If I can make it from ground zero, or sub-zero according to some, they can too. It doesn't mean I'm any saint, I have the same pitfalls as any person; we all have our faults. If we had no faults we wouldn't need priests, or God or anyone. We're all human, everyone gets treated with the same respect, whether they're transvestite, gay, straight, green, blue — we're all humans. I worked in surgery too long," says Morissette of the time he spent as a medic during his Vietnam era military duty. "I know under that thin layer we're all the same.

"It's also a peaceful place, there's lots of trees in memory of people who have died and babies who were born. Nobody's sanctioned, everyone is welcome. It's an oasis away from everything. I have Buddhists and Moslems — everyone — they all come up here to meditate. When you get into the different areas in the gardens, you drown out all of the pressures of the outside; the cars become little waves along the beach."

Eventually everything ends up back at the Pacific Ocean, where we all seemed to have arrived from at sometime or another. As the sun plunges into the Pacific at twilight, many achievements go unsung, many people go unknown, but time's movements in the sky have been tracked by many. Now, as the last of the rays dip into the cool, deep blue, a new millennium begins. The Valley's basic challenges will be the same, only the expectations will be greater in a place that's proven it's up for the challenges that lie ahead.

Photo by Sally Richards

Chapter Five

Photo by Sally Richards

PARTNERS

Building A Greater Silicon Valley
WEBCOR BUILDERS---144
MCLARNEY CONSTRUCTION, INC.---148
SIDEMARK---150
THERMAL MECHANICAL---152
CORNISH & CAREY COMMERCIAL---154
PECK AND HILLER COMPANY---155

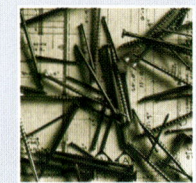

Business & Finance
BDO SEIDMAN---156
GREATER BAY BANCORP---158
TECHNOLOGY CREDIT UNION---160

Manufacturing & Distribution
TREND TECHNOLOGIES---162
TOP LINE ELECTRONICS, CORP.---166
BLISS INDUSTRIES, INC.---168

Quality of Life
PALMER COLLEGE OF CHIROPRACTIC WEST---170
SAN JOSÉ STATE UNIVERSITY---172
VALLEY TRANSPORTATION AUTHORITY---174
HISTORY SAN JOSÉ---176

177 — San Jose Water Company
178 — Symmetry Corporation
179 — University of California, Santa Cruz

Technology

180 — IDT (Integrated Device Technology)
184 — Linear Technology Corporation
188 — AboveNet Communications, Inc.
190 — AITech International Corporation
192 — Altera Corporation
194 — ARCOM
196 — Asanté Technologies, Inc.
198 — Atmel Corporation
200 — ATMI
202 — BEA Systems, Inc.
204 — Cadence Design Systems, Inc.
206 — Cisco Systems, Inc.
208 — Flextronics International
210 — GlobalCenter
212 — Kinetics
214 — PeopleSoft
216 — SEMI
218 — Silicon Image
220 — Winbond

Patron: E-TEK Dynamics

In Silicon Valley

Webcor Builders

TODAY'S WEBCOR BUILDERS IS A UNION OF THE ORIGINAL WEBCOR BUILDERS, INC. AND THE A.J. BALL Construction Company. During the early 1990s, the founding partners of Webcor Builders, Inc., David R. Boyd and Rosser B. Edwards, wanted to make their company more competitive. Well established in San Mateo since 1971, Webcor Builders, Inc. was known for its integrity, quality product, experienced staff and innovative concrete techniques. Webcor had a solid but low-profile reputation. Enter Andrew J. Ball who had left his position as executive vice president of Koll Construction to start his own firm in 1992. Opting to get back into the design and construction process, Ball brought with him corporate campus, hotel and large-project construction experience as well as presentation marketing expertise, but few employees. The pieces of the two puzzles fit and when the companies merged in 1994, Ball became president of Webcor Builders.

After the merger, President Ball created a marketing strategy that included freshening the company logo, designing a new brochure and presenting the Webcor story to the high-tech industry. Local, post-merger clients were Electronic Arts, ATMEL Corporation, Symantec and Electronics for Imaging. Webcor also continued to serve as the general contractor for Oracle Corporation's World Headquarters in Redwood Shores, which Ball can see from his own office window. These local companies continued Webcor's presence in the community and by 1995 secured its position as a builder for corporate campuses and high-tech companies. Webcor's other specialties include speculative commercial office, high-density residential, hotels, large parking structures, renovation and seismic upgrades, interior construction and tenant improvement.

Adobe Plaza, San Jose — an 11-story, 170,000-square-foot office building with over five levels of parking constructed for Adobe Systems
© *Mert Carpenter*

Symantec CC5, Cupertino — a four-story, 143,000-square-foot building and three-level, 150,000-square-foot underground parking structure
© *Mert Carpenter*

The merger was definitely a turning point for Webcor, says Richard A. Lamb, senior vice president and 15-year veteran of the company. "The merger was the key to growth as well as a solution to managing the volume of annual growth that went from $94 million to $526 million, and a salaried staff that has grown from 60 to 150," he says. A separate Interior Construction Group (ICG) was formed to manage the increased demand for project build-out — completing building interiors to a move-in condition. This is a service the company had always offered, but never promoted. Ball and his executives and staff are proud that 80 percent of their business is from repeat clients.

Today, clients will often seek out Webcor to build their buildings, and all

Oracle World Headquarters, Redwood Shores — six office buildings (structural steel with exterior precast panels and glass skin), a power station, a fitness center, a conference center and four parking garages
© *Steve Whittaker*

Electronic Arts Campus, Redwood City — an eight-story building, six-story building, commons building, and 900-car garage, including full interior build out by Webcor ICG
© 1998 Mert Carpenter

Oracle 400, Oracle Corporation, Redwood Shores — 124-story, 294,000-square-foot office under construction
© Jane Lidz

of Webcor's work is negotiated rather than the traditional hard bid. In the traditional bid process, the design and plans are completed before the bid, limiting the range of value and services that a builder can offer. However, clients who desire and appreciate a team approach from a project's inception include Webcor early in the process. This allows Webcor to help the owner assemble the best team of professionals possible, including the architects, engineers, subcontractors and vendors. During the early design phase, a Webcor principal works with the team and uses its experience and preconstruction expertise to offer the owner value engineering and alternative options regarding site use, materials, building systems, equipment, structure, design details, construction methods, cost factors and schedules. Through this process, Webcor can save the owner time and money. This process also allows the team to establish a forum for open communication, which helps to avoid potential problems on a project. Webcor can then build a project of superior quality and provide the owner with the best value for the construction dollar. Webcor's No. 1 goal is client satisfaction, which is reflected in its mission statement. Ball underscores Webcor's commitment to quality, value and client satisfaction by giving each employee a framed copy of Webcor's mission statement and list of core values.

On each Webcor project, there is an experienced project manager who oversees the project and acts as a liaison with outside personnel. In addition, clients are guaranteed that a Webcor principal is active throughout the project. Ball says that as most companies grow, the people with the most direct building experience and knowledge are frequently pulled away to administrative responsibilities. Once contracts are signed, projects are handed off to much less experienced employees. Conversely, Ball's approach is to provide value through active participation in each project and the Webcor principal who works on a project is selected on the basis of his expertise.

Highly regarded for its concrete expertise, Webcor is the only company in the Bay Area that casts building panels at the job site — most companies subcontract the work. White concrete, which Webcor casts, requires attention to the exact mix of fine and coarse aggregates and other ingredients to meet specific standards so that the concrete has a consistent color and dries without cracks. Because Webcor produces such a high-quality product, other general contractors now use Webcor as a concrete subcontractor.

Methods for mixing concrete have not changed greatly over the years, but technology and how technology is used in building have evolved substantially. As Webcor continues to build high-tech corporate campuses, tech savvy is very important. Technology companies expect their builders to be on top of software and hardware developments. Webcor has a project network, ProjectNet, an Internet-based tool for sharing information, documents and drawings among the project team. Every project's project manager and superintendent have individual laptop computers. On-site servers and T1 lines link all job sites together. And project managers, field superintendents and project engineers use hand-held electronic devices to record notes and keep track of contacts and schedules. "It used to be hammer and nails; now it's laptops and Palm Pilots," Lamb says. An in-house server, e-mail and high-tech devices assist Webcor in improving real time communication and adapting to client's needs.

"Technology changes, and with it, the needs of clients change," says John C. Kerley, senior vice president who worked with Ball at A.J. Construction and Koll Construction. Client needs and satisfaction determine what Webcor offers, and when clients need to bring their products to market faster, Webcor has to anticipate potential issues and resolve problems faster. The needs of the end user, or the client's corporate employee, are also important and changing, and this technology has a direct impact on Webcor. According to Ball, Webcor takes a proactive role in research and development of the high-tech business environment. It is one of a dozen companies — but the only contractor — who recommends research topics and issues to the Center for the Built Environment (CBE), a Berkeley research and development organization that conducts studies and publishes its findings. Studies in occupant comfort level include comparing electronic systems as well as the effects of air quality, window coverings, partitions and music on employees. Webcor gains technical knowledge from its initial homework when submitting a proposal to CBE. As a contributor to the research project, Webcor also gets a first look at the findings. "By the time the results are widely published, Webcor will have a good understanding of how to apply the findings to its clients' needs," Ball says.

President Ball clearly enjoys constructing buildings and telling the Webcor story; he also enjoys building a solid future for Webcor. "The future belongs to companies who can move fast, adapt and be flexible," he says. "A company has to be innovative and take risks, but also be smart about its moves. Webcor has diversified, but hasn't spread itself thin." He finds no reason to spread Webcor around the country or add specialties simply because they are on a hot market list. The company's current and future growth are based on the foundation of past successes and an extension of existing strengths — Webcor's Interior Construction Group is one example. New building trends can be inconsistent, but improvements and remodels are a steady $8 billion business in California. In its two years as a separate division, ICG grew from $12 million to $90 million in contracts.

Webcor has grown extensively since the merger, and Ball believes the company has reached an optimal number of employees in which to operate successfully. He and his executive staff looked at ways to create a dynamic, consistent work environment. Working with a corporate consultant, Webcor restructured and reorganized its corporate structure. Policies and procedures were put into place and job descriptions and expected skills were defined. Reviews are used to identify areas in which an employee can improve; then effective training takes place. A mentoring program, in which each employee of eight years or more is a mentor to up to five other employees, helps to support the company's team system. Ball believes that matching employees and mentors with different personalities and styles helps everyone to work on communication skills and grow. Webcor also offers stock ownership to key individuals within the company, creating a pride of ownership among the staff. Vice President Glenn P. Gabel, who has worked at several privately held Silicon Valley companies, says Webcor is unique in its approach to offering stock. "The ownership of the company is shared, creating a different mindset and very low turnover."

The employees at Webcor believe in giving back to their community. The company contributes to and participates in many community projects, including Habitat for Humanity, the Boy Scouts and the Ronald McDonald House. In 1999, for the ninth year, Webcor participated in Christmas in April, a very popular companywide activity. Over 500 Webcor executives, staff, subcontractors and local volunteers swept, cleaned, painted and renovated Capuchino High School in San Bruno. "Christmas in April has become a tradition for Webcor and the direct community contact is important," says ICG director Joe Hansen.

Treating people well — clients, employees and members of the community — is an essential ingredient to Webcor's philosophy, and one that has brought success to its projects throughout the Silicon Valley.

Bayshore Christian Ministries, East Palo Alto — Webcor workers are present at Bayshore's barn raising! Webcor donated its services to erect the frame of this two-story, wood-framed building in only one day! Bayshore Christian Ministries is a community service organization that provides tutoring and mentoring for underprivileged children.
© *Mert Carpenter*

Electronics for Imaging, Foster City — a 10-story, 300,000-square-foot building, including both shell construction and interior build out by Webcor ICG
© *1999 Russell Abraham*

McLarney Construction, Inc.

"I REALLY ENJOY THE ENTIRE CONSTRUCTION EXPERIENCE," STATES KEVIN MCLARNEY, PRESIDENT OF McLarney Construction. "To me, the construction process is more than just structures and building materials; it is a passionate commitment by our entire team to meet the needs of our customers. By team, I am referring to McLarney Construction, our subcontractors and, of course, our customers. As a result of committed teamwork, trusting relationships evolve, and it is these relationships that make the entire process so enjoyable."

Since 1987 when Kevin McLarney first began the company's operations, teamwork has been the key to McLarney Construction's ongoing success. Each person involved, from the tradesman of subcontractors to the office personnel of McLarney Construction, feels a sense of ownership in the successful completion of each project. Kevin has surrounded himself with a community of professionals who mirror his passion for "a job well done." Each person is expected to contribute his or her best and does just that. From architectural services, design build, value engineering and scheduling to the actual driving of screws, painting of walls and laying of carpet, success is dependent upon the success of each person. Project challenges are met with enthusiasm and the entire team benefits from the experiences and problem-solving skills of each individual. The use of the latest technology — cell phones, laptop computers and ricochet modems — links team members between the field and the office. Weekly job site meetings and safety meetings serve to bring everyone together to address specific challenges of each project and industry-related issues.

Kevin McLarney enjoys every aspect of the construction process. Not content to stand on the sidelines coaching his people toward project success, he prefers the role of team captain — getting on the field, getting muddy alongside his teammates. His involvement with each project begins during the initial consultation with clients, really listening to their building expectations and goals. He understands how important it is to select and assemble the right project management team. Project success and client satisfaction depend upon it. Weekly in-house meetings serve to keep McLarney abreast of each project and new client change requests that affect project scope or scheduling.

However, it is during the punch list walk-through that Kevin's commitment to each project really pays off. The punch list walk occurs at the near completion of a project, when the customer and McLarney Construction walk the project site together looking for deficiencies. Items are documented and marked with blue masking tape. When Kevin first began the walk-through review in the late 80s he made certain his superintendents understood his expectations. His tape markings were brutal. In fact, there was friendly banter among McLarney's project team about Kevin's blue fetish. Today, his superintendents pride themselves on the rare appearance of the dreaded blue tape.

McLarney Construction's office reflects McLarney's desire to nurture an environment of teamwork and productivity. He has the ultimate "open door policy" in that his office has no door. It is an open, dynamic and sometimes boisterous environment. It is common for team members to yell out questions, advice, comments or even jokes across the office. The overall feeling in-house is one of community and family — people working together toward a common goal — as opposed to individuals working in isolation.

While the general theme of the office décor is nautical, it is by no means limited to the sea. McLarney feels that

Kevin McLarney, President of McLarney Construction

his group's creativity is enhanced by a sense of adventure infused into the design of the office. An antique motorcycle shares space with the ship's wheel from a U.S. presidential yacht named after the *Mayflower*. Antique guns contrast sharply with whaling harpoons and telescopes. Throughout the office, walls are lined with photographs of McLarney Construction projects, many of which include testimonials from satisfied clients.

McLarney Construction is a solid presence in the Silicon Valley. As the company undertakes and completes different types of projects (corporate, class "A" T.I.s, bio tech/high tech and retail), the entire team continues to strengthen its reputation for producing a quality construction product on time and within budget. The end products often exceed the expectations of the client. Brokers, property managers, facility managers and tenants look to McLarney Construction as a primary construction resource. Customers recognize McLarney's high degree of ethics, thoroughness, responsiveness and sensitivity to costs. McLarney Construction is ranked among the Top 25 commercial general contractors by the *Business Journal of the Silicon Valley*.

"A good construction process often begins with, and should result in, a trusting relationship." Repeat customers comprise 95 percent of McLarney Construction's yearly volume (AboveNet, Catellus, Hewlett-Packard, Spieker Properties, Insignia/ESG and Oracle). Customers experience McLarney's teamwork, execution and strong customer focus firsthand. The team brings various resources to the table to assist both tenant and owner. Clients appreciate the entire company's ability to walk them through the construction process. Constant communication keeps everyone up to date and prevents unpleasant surprises. McLarney Construction is proud of the fact that it does not have a sales department — its repeat clientele has become its sales force.

An important factor in maintaining a strong customer relationship is McLarney's ability to provide current information about changes affecting the construction industry. On-site seminars, digital formats containing company information and client interests, and use of the Web and the Internet are tools employed by McLarney to communicate with both existing and new customers. Communication is an ongoing process with every client and continues long after projects become realities.

The enjoyment of the entire construction process — building, communicating, achieving — is certainly the motivating drive behind McLarney Construction.

Interior of McLarney Construction's San Jose office

SideMark

CORPORATE FURNITURE

SILICON VALLEY COMPANIES NO LONGER PROVIDE OFFICE SPACE FOR THEIR EMPLOYEES; THEY CREATE A UNIQUE corporate culture thanks to companies like SideMark. At first glance the textured red fabric chairs, cream snap-together walls, iridescent mobile filing cabinets and oblong wooden tables displayed in SideMark's showroom in Santa Clara appear to be an interior designer's dream. But upon closer examination, the artistic display of furniture is really the culmination of everything that the Silicon Valley has come to represent. SideMark and its birthplace, Silicon Valley, are places where knowledge and creativity combine with vision and invention, style and skill. In these places, function spawns form and need is synonymous with desire.

The speed with which this 15-year-old company grows only seems to accelerate with time. Every year since 1995 SideMark has been named one of *San Jose Business Journal's* "Top 100 Fastest Growing Companies" in the Silicon Valley." Now, with a gross of more than $30 million in sales, it is one of the top 10 furniture dealerships in the Bay Area. The SideMark client list includes Inktomi, Broadvision and Yahoo, one of the largest Internet-related businesses in the world. When SideMark first contracted with Yahoo!, the new company had 50 employees. Today Yahoo! employs nearly 3,500 and is one of the most recognized names on the Internet. In a place that is home to more than 7,000 electronics and software companies and thousands more startups, where 11 new companies are being created every week, Owner and CEO Randall

Horton has found a unique niche in one of the hottest markets in the world.

Things haven't always been so hot for Horton, however. After graduating with an interior design degree from San Jose State University, one of his first jobs was to design arcades that housed coin-operated video games for Atari Corp. In 1984, after the company had slumped from 10,000 employees to 158, Horton was laid off by the Sunnyvale corporation. He decided to start his own business out of his studio apartment in Palo Alto and contacted everyone he could think of in an effort to drum up clientele. His savings shrunk to $500 before IntelliGenetics Inc. gave him a chance.

Engineers at IntelliGenetics, a 4-year-old software developer in Mountain View, took him to the Palo Alto Research Center (PARC) to provide a model of what they wanted in their own office environment. Beanbag chairs, white boards on the walls and a late-night cafeteria imbued PARC with the feel of college dorm life. To the

Broadvision's open office environment in Redwood City, California
©1999 Steve Whittaker

Broadvision's lobby at headquarters in Redwood City, California
©1999 Steve Whittaker

recent Stanford graduates this was "a computer engineer's heaven." During the ride back from PARC, Horton asked the engineers what they needed in their own space. At the top of their wish list were items like a secure place to store their state-of-the-art bicycles while they worked. Some wanted to bring their dogs with them to work. They wanted food to be available around the clock and, of course, they wanted beanbag chairs.

Horton did better than beanbags. He scoured trade shows and catalogs to find companies that manufactured furniture that didn't just make a fashion statement, but addressed the very specific needs of the highly specialized market in the Silicon Valley. During his search, Horton discovered Teknion, a Canadian-based company specializing in maximum-flexibility furniture designed with high-capacity data and electrical elements in mind. "It's like a Lego system for the office," Horton explains. With the fast-paced growth potential in the Silicon Valley, this type of modular system provided the flexibility, fashion and function he was looking for in a product line. Now, SideMark deals with more than 300 vendors to provide for their expanding clientele, but Teknion still remains one of their leading suppliers.

Images which may have scared investors on Wall Street in the 80s now work as a marketing scheme to attract prospective investors, clients and employees. With the spawning of the "dot-com" companies, or Internet-related companies, the evolution of creating office culture is more important than ever before. The flashy purple-and-yellow color scheme throughout the offices at Yahoo! is just one example of what Horton describes as the "Wow!" factor, which sells a new culture that defies conservative corporate mentality. "Companies like Yahoo! aren't afraid to step out of the box," Horton explains. SideMark worked with Yahoo!'s architect to incorporate the vivid purple-and-yellow of Yahoo!'s graphics into the open office environment. The result has so enamored some employees that a few of them have covered their own automobiles with the trademark purple-and-yellow decals.

Yahoo! lobby at headquarters in Sunnyvale, California
©1999 Steve Whittaker

Standing in his storeroom of color, fabric and texture samples, it is clear that Horton has no problem meeting the needs of the Gen-Xers as well as his more conservative clients. More than 10 floor-to-ceiling bookcases house hundreds of books that offer countless fabric, material and color samples. Color options range from Suburban Shadows to La Plume. Spa Sea is a counter top option that looks like a swimming pool.

"You want a black leather chair?" He laughs, "Here, look through this 25-page book and try to decide what shade of black, what type of material, what texture you want on the material, and then I have a warehouse out back to help you decide what features you want on it!" Luckily for Horton, his 20 account and project managers are usually the ones faced with the daunting chore of narrowing the vast selection down to a few key choices.

Horton's black chair sits in front of a window that looks out across the industrial sprawl of a northeastern corner of the Silicon Valley. It has a built-in massage unit and he calls it the "Executive model." When asked if he ever thought he would be at the helm of something so big, he just laughs, "I don't know. I just took it one step at time."

Yahoo! Teknion workstation in Sunnyvale, California
©1999 Steve Whittaker

Thermal Mechanical

Richard Rood, President

RICHARD ROOD LIKED HIS FIRST JOB. ABOARD A NAVY HOSPITAL SHIP IN THE MID-50S, HE WAS TRAINED TO keep the medical supply refrigerators and air-conditioning systems cold. But he never thought this assignment would lead him to a career creating the ideal environments for space research, atomic energy and cutting-edge computer technology.

Returning to the mainland during a time when only hospitals and the very wealthy had air-conditioners, Rood started out by installing and maintaining refrigeration systems in grocery stores. He soon became a piping contractor when his customers needed gas lines installed to supply their updated equipment. In 1968 Rood founded Thermal Mechanical, a company that today specializes in air-conditioning, refrigeration, mechanical piping and plumbing. His timing couldn't have been better. At the birth of the electronic industry in the Silicon Valley, a year prior to NASA putting a man on the moon, the race for technology was just heating up.

It wasn't long before NASA called and asked if Thermal Mechanical could build a low-temperature test chamber. Rood replied, "No problem." Next NASA wanted an altitude chamber. "We can do that too," he said. When Stanford Linear Accelerator (SLAC) needed a Chiller Application, a specialized cold trap used when splitting atoms, Thermal Mechanical was there to build it.

And that was only the beginning. In the late 70s, Applied Materials, IBM and Intel soon became clients when Thermal Mechanical began building clean rooms and wafer fabrication rooms. Both of these rooms provide the necessary environment required during the intricate transformation of quartz crystals into silicon wafers and then to microchips.

Throughout the 80s Thermal Mechanical manufactured burn-in rooms (rooms heated up to more than 100 degrees to show an accelerated rate of failure) for companies that were testing and developing hardware. Biotech laboratories were soon contacting Thermal Mechanical to address their specialized high-purity piping and air-conditioning needs.

Thirty years after the company's inception Rood looks back on its history and says, "From the technology standpoint alone it has been fantastic! Our company has evolved out of customer need." If this rich history gives any indication of the future, Thermal Mechanical is in for quite a ride.

The company now sits on two acres in the heart of the Silicon Valley with more than 120 employees and an in-house sheet metal fabrication department. Rood takes pride in the variety of clients his company has worked with throughout the Bay Area. He lists high-tech industries, schools, hospitals, offices, churches and department stores as sites of just a few of his satisfied customers.

"Air-conditioning is still the meat of our business," Rood explains while seated comfortably behind his large, dark wood desk. With the continuing expansion in the business office sector in Silicon Valley, this is not surprising. The latest wave of successful Silicon Valley entrepreneurs probably don't want outdated air-conditioning systems in their state-of-the-art office space. But Rood, who modestly

> IN 1968 Rood founded Thermal Mechanical, a company that today specializes in air-conditioning, refrigeration, mechanical piping and plumbing.

shrugs off what he calls "high-tech gadgetry" in his own office and says he would rather get a letter than an e-mail any day, says that air-conditioning has already witnessed its evolution.

"Compressors haven't changed substantially in the last 50 years. The whole refrigeration cycle is still the same. Components have changed, but the basic principles have not."

What has changed is the application of air-conditioning. Some of the large Internet server sites may have 10,000 square feet of electronic equipment. "Just imagine rows and rows of machines," Rood says sweeping his hand through the air to demonstrate. "Think of it like a light bulb. Once you turn it on, it starts to generate heat. Our job is to remove the heat to prevent the system from burning up. The computer system cannot operate without cooling."

One desktop computer generates roughly the same amount of heat as a working person, according to Rood, but many of the larger companies also have huge back-up systems, generators and battery back-ups that contribute to the cooling needs. Combine that with the filtration needs of a clean room that requires massive air flow through heavy-duty filters usually placed in the ceiling, and it becomes obvious that air-conditioning is an integral part of modern technological advancement. This is the key link that almost guarantees Thermal Mechanical's future success. Rood believes that the rapid pace of development will continue in the Silicon Valley and that the benefits of this development will impact the world.

Rood, however, is content just keeping things cool. His methods are successful. Nearly 90 percent of his clients are repeat customers, and despite efforts to control growth, his company continues to increase business at a steady rate of 8 to 10 percent every year. His employees seem satisfied as well. The average length of employment at Thermal Mechanical is 15 years. Rigorous apprenticeship training programs and ongoing classes ensure each service technician and installer has knowledge of the latest industry standards and techniques. The company provides a full range of capabilities including project management, engineering, construction, fabrication, and testing and maintenance. Thermal Mechanical's fleet of bright yellow trucks is a common sight around the Bay Area.

Chilled water piping system for a growing Internet Service Provider — designed and built by Thermal Mechanical

Diesel fuel system with failsafe electronic controls — designed and built by Thermal Mechanical

Despite the ongoing success of his company, Rood maintains a genuine appreciation from where that first job and a lot of hard work have brought him. Rather than trying to impress clients with modern furniture, expensive art or advanced technology in his office, Rood prefers his visitors admire the carved wooden elephant head with a singular white tusk he bought on a trip to Bangkok with his wife. Or the replica of the hunting rifle his staff presented to him on his birthday. Or the model airplane enclosed in glass next to the door that resembles the real one he likes to pilot. His desk is free of everything except good old-fashioned paper and pens. Indeed, it is within his ability to balance the knowledge of the past and respect for the future that Thermal Mechanical has found a firm foundation to build upon.

Cornish & Carey Commercial

IT IS SAID THAT SOME PEOPLE FEEL EARTHQUAKES BEFORE THEY HAPPEN. SUCH IS THE CASE OF THE FOUNDERS of Cornish & Carey Commercial (C&C). Long before the Silicon Valley technology boom registered on the country's economic Richter scale, they were prepared. Since its inception in 1935, Cornish & Carey has grown to become the largest commercial real estate company in the Silicon Valley, totaling over $3 billion in real estate transactions in 1999. It has accomplished this by having the vision to see change coming and the talent to embrace it.

Cornish & Carey Commercial began in Palo Alto when George E. "Pat" Carey and Herbert J. Cornish merged their two real estate businesses. This is the building that now houses the Palo Alto offices.

George E. "Pat" Carey, whose family came to the valley in the 1880s, and Herbert J. Cornish, a valley native, each had real estate businesses in Palo Alto when they decided to merge the two in 1935. From the very beginning, their combined experience in the area enabled the company to shake up the competition. Cornish & Carey's growth, throughout its 60-plus years in business, has paralleled that of the Silicon Valley, as it played a key role in the area's development.

As technology took off in Silicon Valley, Cornish & Carey developed strong relationships with some of the industry's key players. Beginning with Hewlett-Packard and Varian, C&C has also served Intel, Apple Computer, Xerox, IBM, Netscape, Metropolitan Life and 3-Com, among others, and many of them through decades of growth. Over a seven-year period, one high-tech company started with 5,000 square feet and today leases or owns over 1 million square feet — all through Cornish & Carey and its professionals.

C&C provides service to properties of all sizes and types including facilities for biotech, semiconductor, computer electronics and defense-related companies. Its services include consulting, site location, leasing and sales, property analysis, investment sales and market research — offering expertise throughout the duration of each transaction. The company ranks as the largest independent commercial brokerage firm in the Pacific region of the United States, and is in the top 10 nationally. Its affiliation with ONCOR International enables it to serve companies with international interests as well. C&C is owned and operated by its eight top executives and employs more than 200 sales agents and support staff in six offices.

Scott Carey, nephew of founder Pat Carey and a former mayor of Palo Alto, is the chairman of Cornish & Carey Commercial. He says building relationships in the community provides the knowledge and reputation to do a job successfully. Those relationships, together with an unmatched expertise in solving real estate needs, an expertise acquired by over a half a century of business, provide a real "value added" service to its clients. And, he says, by providing service of value, the company's clients come back:

"For more than a decade, this division (Westinghouse) has employed the real estate services of Cornish & Carey. Whether we need 5,000 square feet for a temporary office, a 105,000-square-foot warehouse on a long-term lease, or the master planning of our 80-acre site, Cornish & Carey has been an innovative, creative partner every step of the way."

— Sharon Barney, Real Estate Services Manager

C&C's strategic plan is to expand into the San Francisco and East Bay market, and to broaden its advisory and investment services. It will continue to build relationships with technology companies in all phases of their growth, creating "partnerships" for the future.

Peck and Hiller Company

Located in East Palo Alto in the heart of the valley, Peck and Hiller Company, a leader in structural concrete contracting, recently celebrated its first half-century of operation. The company was founded in 1949 by Robert Hiller and Earl Paddack. Russ Peck and Tom O'Connor joined the company in 1961, bought Hiller's and Paddack's interests in 1964, and became president and vice president, respectively.

Originally a steelforming company that used steel forms exclusively, Peck and Hiller expanded its scope first from steel forms to all formwork systems available, and then to concrete package work. Peck and Hiller offers a full range of concrete services, primarily creating partial or complete formwork for structural concrete. The company also offers a complete structural concrete package, handling all the formwork, including footings and placement of all structural concrete, and subcontracting the reinforcing steel and concrete placement.

Peck and Hiller has earned its rating as a valuable asset to the construction industry by working closely with prime contractors in cost analysis, developing budgets and schedules, and creating the best building scheme in a project's planning stage. The company has built projects throughout the United States, from Connecticut to Washington, D.C. to Louisiana, where the company constructed 1 million square feet of formwork for the Louisiana Superdome. Recent West Coast projects range from parking structures to hotels, high-rise office towers, college buildings and residential condominiums, to convention centers and major sports arenas.

The company's largest formwork job is the $6-7 million project for the San Jose Convention Center, which includes two underground parking levels. Other multistory parking projects include two San Francisco International Airport parking garages; the Pier 39 parking structure in San Francisco; and the Galleria parking structure in Sherman Oaks. The company provided the arch formwork for San Francisco's George Moscone Convention & Underground Exposition Center. The company's largest concrete package project is the Apple Computer campus in Cupertino, with a contract value of $10 million.

Peck and Hiller's long list of formwork projects for hotels includes the Red Lion Inn in San Jose; the Palo Alto Hotel; the Marriott Hotels in Santa Clara and Anaheim; and the Holiday Inn in Ontario. Peck and Hiller provided the formwork for two 20-story buildings, including the luxury L'Elysee Condominiums in Los Angeles. San Francisco residential projects include the Opera Plaza Condominiums and Mei Lun Yuen Chinatown Housing.

Several Bay Area educational facilities have the Peck and Hiller mark on them: the Moffett Undergraduate Library building at UC Berkeley; the San Jose State University Library building; and at Stanford University, the Earth Science Laboratory and classroom building, main library building, Roscoe Maples Basketball Pavilion, Stanford Business School classroom building, and the Center for Clinical Scientific Research.

Peck and Hiller is a union contractor employing an average of 150 construction workers at any given time. Employees are involved in community outreach activities such as Habitat for Humanity and Christmas in April. The company owners are active in construction organizations, concerned with promoting job growth and health and safety issues, and provide many workers with on-the-job training, enabling them to learn marketable skills. Looking to the future, the company plans to re-enter the Southern California market in 2000.

Peck and Hiller provided the arch formwork for San Francisco's George Moscone Convention Center.

BDO Seidman

AS A MIDDLE-MARKET BUSINESSPERSON OR ENTREPRENEUR, IT IS NOT UNUSUAL TO PONDER DEATH AND taxes with an equal amount of dismay. In addition to handling IRS audits, communications and filings with the SEC, and determining a company's overall tax compliance, a trusted accountant can be an ally to business growth and productivity.

Such is the mission of BDO Seidman, LLP in San Jose. This accounting and consulting firm excels in client-based strategic financial solutions with a limitless variety of services designed to build business and nurture growing companies well into the next phase of their development. BDO Seidman serves clients through more than 100 locations across the United States. As a member firm of BDO International, BDO Seidman can leverage a global network of resources to serve clients abroad through more than 500 member firm offices in over 90 countries with approximately 19,000 partners and staff.

BDO Seidman's San Jose office opened in 1990, with a focus on emerging growth companies. Just 10 years ago, the San Jose practice was facing the same challenges that many of its startup clients are experiencing today. The four partners, Bob Lanz, Wynne Meredith, Isabel Chiu and Brian Cardozo, who each have spent between 20 and 30 years as accountants, haven't forgotten what it feels like to carve a niche in today's competitive business environment. With growth at BDO Seidman exceeding more than 25 percent each year since 1991, the partners at BDO have discovered what it takes to succeed.

BDO's team of experienced accountants work more like business advisors as their services reach far beyond that of a typical single-service accountant. "We get closer to our clients than most accounting firms," says Partner Brian Cardozo, seated in one of BDO's conference rooms overlooking the swimming pool at the Fairmont Hotel in

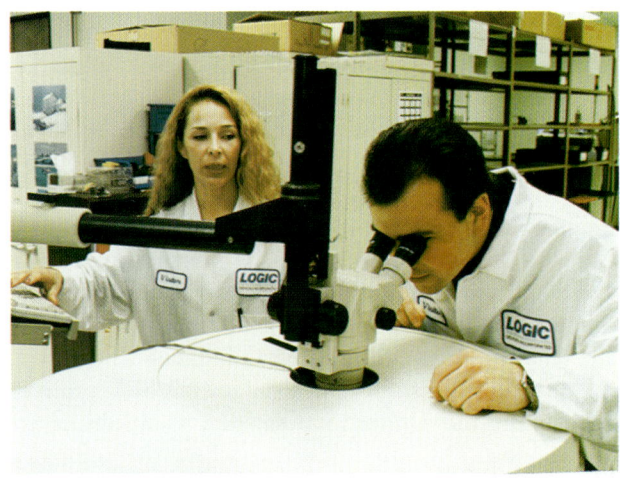

Audit staff accountant Chris Cunningham at client site — Logic Devices, Inc.

Staff training session in the main conference room of the San Jose office: (left to right) Carrie Lo, Andy Chao, Debbie Scherer, William "Rocky" McDonald, Robert Lanz, Brian Cardozo, Susanna Winchell, Chris Cunningham and Robert Akerblom

downtown San Jose. "We make sure they get on track and stay on track by building relationships that last years."

Their business isn't just about numbers; it's about people. Through its vast network of business professionals, BDO strives to connect the right people to the right services. BDO provides clients with access to seasoned professionals who offer a full range of financial services including lines of credit, lease lines, stock option programs and e-commerce solutions.

Mergers and acquisitions are an integral part of Silicon Valley business, and the partners at BDO consider themselves well versed on the subject. They can oversee the entire transaction or just advise clients on the most advantageous route through the harrowing course of relinquishing shares during the often stressful rounds of financing. Always on the lookout for enhancing business opportunity, the partners at BDO won't hesitate to recommend strategic partnerships with both domestic and international vendors.

BDO accountants will even fill the role of a resource broker by offering in-house expertise to companies that either can't afford or don't usually have the need for their own internal financial team. For example, a company in Sacramento engaged a BDO partner to be its chief financial officer during the negotiations of a sophisticated transaction. Afterward, BDO implemented a marketing

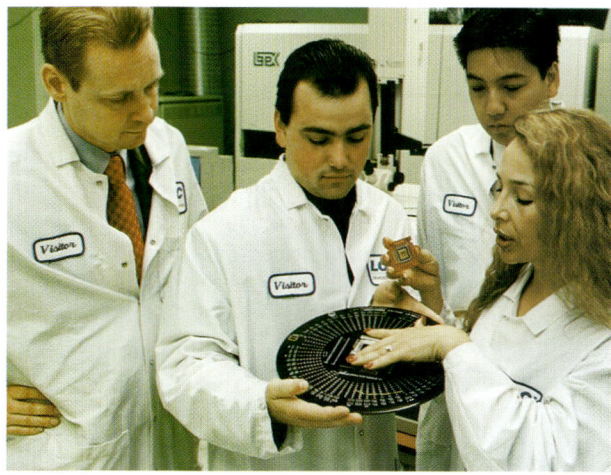

and advertising campaign that helped grow the business from $3 million to more than $11 million in just three years. Serving as the company's representative, BDO then identified a strategic buyer for the company and negotiated the transaction through to consummation of the sale.

BDO's client services span the full gamut of financial needs, right down to the office equipment. Setting up migration paths for software and hardware systems in growing businesses is another detail they do not overlook. "Just because a company can't start out with a $100,000 accounting package doesn't mean they can't set themselves up to grow into one when they need it," says Cardozo.

This kind of support has helped fuel BDO Seidman's growth and presence as one of the nation's largest accounting and consulting firms. With such a vast network of business line partners, the company can pull from within its ranks the resources needed to meet client demands. BDO takes a proactive step by providing clients a free-of-charge yearly visit in which they discuss the latest developments in company growth and offer direction

Partners, managers and staff of the Silicon Valley Practice during a recent presentation on firm initiatives and practice updates

before financial mistakes are made that could prove to be costly down the line.

"We aren't afraid to think out of the box. We are creative, aggressive and committed to the projects that we take on and the work that we do," Cardozo explains. The partners at BDO demonstrate their "can-do" attitude in more than just the businesses around San Jose; they are also making a difference in the community. Cardozo is on the board of directors at the Santa Clara and San Mateo County Child Abuse Prevention Center. Partner Bob Lanz is on the board of the San Jose Cleveland Ballet. Isabel Chiu actively participates in the Cancer Society. In addition, the office provides tax and business advisory services for the Junior Achievement, Junior League and the Kuumba Jazz Center.

Their outside interests may be varied, but each of the partners retains a strong commitment to the betterment of their community as a whole. Although they can't change the inevitability of either death or taxes, the people behind BDO Seidman in San Jose are proving that accountants can be strategic business partners for life.

Audit manager and staff during a tour and presentation regarding the client's business and products

(Left to right) Isabel Chiu, Partner in the Milpitas office, Robert Akerblom, Senior Manager, and Robert Lanz, Partner in the San Jose office

Greater Bay Bancorp

THE WALLS RISE NEARLY A HUNDRED FEET TO MEET THE ARCHED PLASTER CEILING, GIVING THE PALO ALTO headquarters of Greater Bay Bancorp (GBB) a venerable old bank feeling. But hidden behind the rather austere surroundings is a lean, smart, aggressive financial institution that has become the dominant independent business bank in the Bay Area, with over $3.2 billion in assets and 19 offices ringing the bay from San Jose north along the Peninsula to San Francisco and in the East Bay to Walnut Creek, as of the end of 1999.

Greater Bay Bancorp (NASDAQ: GBBK) is the holding company formed in 1996 by the merger between Cupertino National Bancorp and Mid-Peninsula Bancorp. The merger gave Cupertino National Bancorp access to the important northern Peninsula market (San Mateo to South San Francisco) while Mid-Peninsula Bancorp gained access to added small-business lending and trust services expertise. More importantly, a new banking strategy and management philosophy was born.

David Kalkbrenner, GBB's President and CEO, was a longtime executive at Crocker National Bank. When Wells Fargo bought Crocker in 1986, Kalkbrenner felt it was time to move on. So when he co-founded Mid-Peninsula Bank the following year, he vowed to keep his employees in place whenever a merger or acquisition occurred. This philosophy, built on empathy and shared experience, tied into a strategy Kalkbrenner and his directors called super community banking.

Kalkbrenner's team realized that commercial bank customers — in particular, the small-business people who operate businesses with annual revenues between $5 million and $100 million — place a high value on doing business with institutions that are locally managed, where financial decisions are not kicked upstairs or sent to a headquarters located in another city or state. He also knew that business customers need an ongoing relationship with a full-service bank.

The super community banking structure that was created to deliver these benefits is a three-pronged approach that includes a strong community focus, a broad line of commercial services and corporate cost savings that are completely transparent to the client. More importantly, perhaps, is the fact that this structure has allowed Greater Bay Bancorp to grow through mergers and acquisitions without sacrificing the relationships and continuity each bank's clients had come to trust.

GBB's strong community focus means subsidiary banks operate as individual community banks. The original bank names are retained, as are board members, bank executives and staff. A model was created for local decision making, giving individual bank officers the authority and flexibility to deliver services based on their understanding of the local business community's needs. These officers, in turn, participate in their local community's activities, thereby opening an informal communications and knowledge-trading channel.

The second part of the super community bank strategy was built around products and services. An independent, local bank is hamstrung by the need for specialized personnel and a supporting infrastructure in addition to its cash limitations. But the multibank holding company structure gives Greater Bay Bancorp the resources to offer a broad array of financial products, customized as needed, at very competitive costs. For instance, where most community banks limit loans to under $5 million, GBB clients generally enjoy a limit in excess of $25 million. In some cases, through the company's Corporate Finance Group, syndicated loans up to $250 million are available.

Members of the Executive Management Committee of Greater Bay Bancorp are: (seated left to right) Richard M. Kahler, Murray B. Dey, David L. Kalkbrenner, Frank M. Bartaldo Jr., Kimberly S. Burgess, Steven C. Smith, James R. Woolwine and Linda M. Iannone; (standing left to right) David R. Hood, Gregg A. Johnson, Susan K. Black, Kenneth D. Brenner, Mark F. Doiron, Colleen G. Carlsted, Shawn E. Saunders and Cheryl G. Howell. Not pictured are Nord Hastings and Jim Mayer.

Lastly, Kalkbrenner and his managers knew that in order to grow, GBB had to prosper financially, and to encourage this a unique money-saving system was devised. Simply, administrative support services like investment management, data processing, accounting and functions that support specialized products were centralized and consolidated. This freed the individual banks to concentrate on their clients' needs and to maintain a high degree of personal attention. It also provided a readily available pool of financial specialists with the expertise to help clients navigate through complex areas like international banking, trust services and cash management.

The company established a group of subsidiaries and divisions — the Cash Management Group is one example — so that any client could walk into any bank and gain access to what had traditionally been big-bank services and expertise. These divisions include Pacific Business Funding to provide asset-based financing to companies that might not qualify for traditional bank financing; Greater Bay Trust Company to provide trust and fiduciary services, asset management and investment services; the Venture Banking Group to support emerging and growth-stage technology companies; the International Banking Division to provide local, professional banking service to companies doing business overseas; and the Greater Bay Bank SBA Lending Group (a preferred SBA lender) to offer expedited credit decisions and rapid turnaround to small businesses seeking SBA loans.

One additional entity is particularly close to the hearts of GBB officers. The Greater Bay Bancorp Foundation was formed in early 1998 to provide the means by which the company, its officers and directors could give something back to their local communities. Focusing on local education, health and economic initiatives, the foundation has already provided over $1.5 million in grants to nonprofit community organizations.

The proof of any concept, of course, is in the performance. And by most measures, Greater Bay's vision has worked very well. With two recent acquisitions — Mt. Diablo National Bank and Bay Bank of Commerce —

Standing are: John M. Gatto, Co-Chairman of the Board, and David L. Kalkbrenner, President and Chief Executive Officer. Seated is: Duncan L. Matteson, Co-Chairman of the Board.

GBB now serves clients throughout the San Francisco Bay region. Annual Revenue growth has been 21.76 percent annually for the last three years and Return on Average Shareholders' Equity has topped 22.5 percent annually as of December 31, 1999. In September 1999, Greater Bay Bancorp was named one of *Fortune* magazine's "100 Fastest Growing Companies." And Greater Bay Bancorp was placed on the Nasdaq Financial-100 index, an endorsement of Greater Bay Bancorp's strong financial performance.

But it's not all dollars and cents at GBB. Financial success notwithstanding, Kalkbrenner knows his bank's success rests on delivering personal attention and fast turnaround on a broad range of services. This means the crucial ingredient is Greater Bay Bancorp's people. Management looks for intelligence, energy and integrity, people who understand that banking is a service business. And he makes certain that GBB's people know they are managing not just an account, but a relationship.

Technology Credit Union

BECAUSE TECHNOLOGY PROFESSIONALS IN THE SILICON VALLEY EXPECT SO MUCH FROM THEMSELVES, they set equally high standards from those who serve them. With no time to sit still, except to measure how far ahead they are of the competition, they expect their financial institutions to be as farsighted and innovative as they are. Since 1960, Technology Credit Union has kept up with its highly driven members by offering them a wide range of services targeted directly to their needs. By its charter, the credit union limits its membership to workers and their relatives in the technology industry. As this industry has achieved such conspicuous success, Technology Credit Union, casually known as Tech CU, has prospered as one of the valley's most successful financial services providers.

A credit union is a financial institution organized by a group of people with something in common, through work, education or some other affiliation. While a credit union offers its members many of the same services as a commercial bank, it has a significantly different mission: It is answerable only to its members, not to outside investors. A credit union is owned by its members, and it is run by an unsalaried board of directors that is elected by the membership. Since credit unions are not directed by stockholders, all planning, operations, financial management, services and product development are conducted for the benefit of the members. A credit union

Gloria Debs, Employee Services Manager at Sun Microsystems and an enthusiastic Tech CU member

> **WHILE** more and more of its services continue to be Web-based, Tech CU hasn't forgotten the importance of personal service.

uses its revenues to improve services and to offer competitive rates on loans and deposits. Members of Tech CU know that their financial partner is working solely for their benefit, not to generate dividends for stockholders.

Tech CU was started in 1960 by a small group of employees at Fairchild Camera and Instrument Semiconductor Division in Mountain View. Its founding members were rank-and-file tech workers who planned their fledgling financial enterprise in the company cafeteria, got a boost with a loan from the Lockheed Credit Union and obtained a federal credit union charter. After one year they had 600 members and holdings of $65,000. Within 10 years, membership had grown tenfold and assets had soared to $3.5 million. By the mid-70s Fairchild was no longer in business but the credit union was booming. In 1983 it changed its name to Technology Federal Credit Union, started enlisting other companies and by 1987 had a roster of 25 major employers. By the end of the 1990s, membership in Tech CU exceeded 61,000 and in the last two months of 1999 alone it added 120 new employee groups while staying faithful to its customer base.

People who work on the cutting edge of technology expect the same level of expertise from their money manager, and Tech CU can credit much of its rapid and consistent growth to being able to meet its clients' demands. Today, Tech CU is among the top 1 percent of the country's largest credit unions and is the biggest multiple-group credit union in Silicon Valley serving only technology workers. Its strong commitment to provide its members with growth, safety and stability is matched by its ability to meet demands for speed, accuracy, price advantages and electronic delivery. By the early years of the 21st century, its assets are expected to top $700 million. If Tech CU were a bank, it would be the fifth largest in the region.

However, with their credit union's financial services, Tech CU's members really don't need a bank. They have access to a full range of services in an institution that seeks out inventive solutions and is not limited by tradition and habit. Loan rates are low, savings products produce high yields, service is of the highest quality and the staff is friendly and supportive. Tech CU's members can take advantage of numerous financial tools and services, including checking accounts, payroll direct deposit, overdraft protection, six automated teller networks, VISA CheckCards, electronic funds transfer, and domestic and international wire service.

Technology Credit Union headquarters

Consumer services include loans for new and pre-owned cars, an exclusive auto purchasing network, specialty vehicle and motorcycle loans, signature loans, Guaranteed Auto Protection waivers and Mechanical Breakdown Insurance. Credit services include VISA Classic, Gold and Air Gold cards, Cash Command and Executive Lines of Credit. First-time home buyers can secure a home loan with no down payment, while other home financial services include first and second mortgages, home equity lines of credit, nonowner-occupied mortgage loans and mortgage loan advice. Stock loans and share-secured loans are also available. Savers can open insured savings accounts, IRAs and Flextime Share Certificates, while investors can choose from mutual funds and annuity contracts. Tech CU's other important services include help with life insurance, long-term care insurance, financial planning, estate planning and personalized financial advice.

As part of its commitment to keep its services abreast of advancing technology, Tech CU introduced online trading in the spring of 2000. Tech CU's WebBranch allows its members to conduct some of their financial business from their computers where they can easily transfer funds between accounts, check their balances, verify a loan status, pay bills and search for information on their transactions. To ensure privacy, WebBranch services are protected with the highest encryption standards.

While more and more of its services continue to be Web-based, Tech CU hasn't forgotten the importance of personal service. Currently it offers its members four convenient locations in Silicon Valley and has plans to open one new branch a year through 2004. Tech CU belongs to the ATM CO-OP Network that gives its members access to hundreds of local ATMs and thousands of ATMs nationwide so that, at no charge, members can use another credit union's ATM for both deposits and withdrawals. Membership satisfaction is very high, with customers crediting the organization's terrific service, responsiveness and convenience.

Since its members have little time to worry about their daily financial transactions, Tech CU has two oversight groups. These groups guarantee that the credit union will adhere to the highest operating standards, contributing to its stability and the quality of its management. As a business, Tech CU has a commitment to its employees to foster excellence, promote from within and encourage superior creativity and teamwork in a supportive environment that will empower and enrich them. With four decades of serving Silicon Valley's technology professionals, Tech CU has developed exactly the right approach to support its high-achieving members well into the new century.

The Board of Directors

Trend Technologies

PLASTIC AND METAL HAVE ENDLESSLY INVENTIVE PARTS TO PLAY IN INDUSTRIAL TECHNOLOGY, AND COMPUTERS ARE a primary example. Inside are networks of chips and wires that can't be used until they are encased in a protective outer skin. While intricate electronics are the products of the companies whose names they bear, the critical casings in which they are housed are the work of Trend Technologies. The Silicon Valley-based company makes custom plastic and metal parts for the elite of the electronics giants, and it is one of the top five manufacturers of its kind in the United States and among the top 10 globally. While headquartered in San Jose, Trend has established additional national and worldwide facilities that enable it to design and produce its products in proximity to its high-profile clients. As Trend puts it, the company is a "local source... globally."

Trend has successfully fused the functions of two industries, plastic-injection molding and metal-stamping fabrication. It is one of the few companies with the experience and expertise to integrate inner metal components with their surrounding plastic enclosures. Trend targets Fortune 500 companies, and its clients are primarily the business equipment manufacturers in the computer and electronics fields. Once a client has the idea, it is Trend that rushes it to market, and in a marketplace that won't wait, ramping a product to volume and maintaining on-time delivery are crucial for staying ahead of the competition. To keep pace, the company's volume is extremely high: in a typical month in 1999, Trend manufactured over 230,000 computer chassis for one product line.

Trend presents its clients with a complete package of services, including design and engineering, tool and equipment development, molding, stamping, supplier management, assembly, testing, logistics and quality control. Because Trend's parts are all custom designed and manufactured, the company works closely with its clients, joining forces from the initial concept phase to implement the design. By contributing its experience and expertise, Trend helps its clients plan up front and prepare for all the complexities of design, production and testing. Setting up a successful early strategy eliminates glitches during production and enables Trend's customers to save both time and money.

In today's era of fast-paced manufacturing, products typically have a shelf life of 18 to 36 months, so every year Trend partners with its clients to develop numerous new products. The biggest challenge is to move a product from idea to reality. This process requires Trend to build

One of the buildings at the San Jose home base of Trend Technologies

new tools that will mold and stamp design concepts into real products in the shortest amount of time. Often Trend will build duplicate sets of tools and deploy them globally for its customers to ensure minimal freight and logistics costs and the shortest time to market.

When Trend began in the early 70s, the pace of manufacturing wasn't as fast, nor was technology as highly developed. The company was founded in 1973 by Gino Cavallini and was known as Tool Tech. It started out making plastic injection molds and then moved into manufacturing parts. By 1979 the company had a second division, Trend Plastics, and in 1980 it installed injection-molding presses in its current San Jose facilities. In the early days Atari used Trend Plastics in its manufacturing, and the company then caught the eye of Apple, which hired Trend as its first molder. It was then that Trend moved into making computer enclosures and laptops. As some of Apple's engineers moved on to Cisco Systems and Sun Microsystems, they praised Trend Plastics' capabilities, and soon the company began supplying these electronics firms.

In 1993 Trend Plastics began an expansion program, building a facility in Albuquerque, New Mexico. The following year, the company added an additional site in Colorado Springs, Colorado. In March 1997 Cavallini sold his rapidly expanding company to the Riverside Company, a private investment firm based in New York. Trend Plastics/Tool Tech was renamed Trend Technologies, Inc. The new Riverside management team continued Trend's expansion, creating a new facility in San Diego, California, and acquiring Plasco, Inc. in Longmont, Colorado, and Cam Fran Tool Company in Elk Grove Village, Illinois. The acquisition of Cam Fran expanded Trend's capabilities to include metal stamping. By the end of 1997 Trend had sales of $109 million and ranked No. 32 among North American injection molding companies.

In 1998 Trend took its first step in its global expansion program by acquiring Ballymount Precision Engineering in Dublin, Ireland. Ballymount began operating under the name Trend Technologies Europe Ltd. Its core business was supplying PC manufacturers with metal components, and the acquisition gave Trend a firm foothold in the European market. The companies were a good fit as both had been manufacturing PC equipment for roughly the same number of years, had experienced dramatic growth

Inside the assembly facility at Trend Technologies, San Jose, California

in a very short time and were dedicated to producing the highest-quality products with equally high standards of production. Both prided themselves on their proven ability to adapt to rapidly changing markets and to quickly meet new customer demands. Trend management decided to add plastic capabilities to the two metal manufacturing acquisitions to offer a more complete manufacturing solution under one roof.

During the fourth quarter of 1998 Trend opened still another new plant in Round Rock, Texas. Housed in two buildings and totaling 190,000 square feet, the new facility was Trend's first plant that combined injection molding, metal stamping, mechanical assembly, tooling maintenance and logistics management all under one roof. The facility is located on the outskirts of Austin and is situated there to serve Round Rock-based Dell Computer, one of Trend's primary customers. Being in such close proximity to Dell allows Trend to meet the computer giant's demand for ample parts inventory and quick delivery. By

Trend Technologies' injection molding operations

Products from Trend Technologies' manufacturing facility

integrating so many capabilities at the same plant, Trend is able to keep costs down and create added value for its customer.

Trend knows its business is all about being there for its customers. If a client wants the company to set up shop in Asia, Trend makes plans to be there within six to 12 months. By understanding that a global business also demands a local presence, Trend moves with its customers to serve them quickly and economically. In addition to its many manufacturing locations, the company has distribution locations in St. Petersburg, Florida; Salt Lake City, Utah; Dumbarton, Paisley, Irvine and Livingston, Scotland; Limerick, Ireland; and Penang, Malaysia.

The round-the-clock activity at Trend's San Jose plant represents the company's manufacturing process. The San Jose plant consists of six buildings, comprising 150,000 square feet, and operates 24 hours a day, seven days a week. There are 45 molding machines (it's possible for a Trend facility to consume more than 30,000 pounds of plastic resin in a single day) along with metal stamping, pad printing and assembly lines. The dominant material is PC/ABS along with some pure PC. Most of the presses are Van Dorns, Milacrons and Engels and range from 35 tons to 1,650 tons. While more than 450 people report to work each week, robots are used with 85 percent of the presses. Occasionally a floor operator will intervene, but the employees' primary jobs are to assemble products, box the parts and maintain flow. Two labs constantly oversee quality, while the entire production process on the floor is monitored by 35 PCs that track time and yield, molding and shipping.

Building new molds is the job of Tool Tech, which is one of Trend's tooling divisions. Among its 55 employees are 11 mold makers who produce approximately 20 new molds a month, most of which are very complicated and highly detailed. Every new design is divided into its functional parts, and all the components are built by different tooling departments. By adopting synchronized construction and installing a paperless tracking system, the Tool Tech division dramatically reduced Trend's production lead time. In 1994, lead time was 14 to 16 weeks. Five years later it was fewer than 10. Undoubtedly future engineering innovations will shorten that time again.

Despite its rapid expansion, Trend knows well that there's no resting on its many accomplishments. The competition is too keen for complacency. Therefore, through 1998 and 1999, the company spent $35 million to address the needs of its customers and to grow what it calls its "global footprint" by expanding its integration capabilities and enhancing its presence in Asia, Europe and Mexico. It also initiated a strategy to greatly increase the company's revenues. In 1997, Trend had formulated a plan to grow to a $500 million company. However, the market moved too quickly, and Trend followed suit by setting its goal to be a billion-dollar player within the first years of the new century. Since then it's been moving fast to meet its target. In 1998 the company employed 1,800 people and had revenues of $200 million. In 1999, revenues climbed to $300 million and 800 more employees were added to the company's payroll. Projections for 2000 included $450 million in revenue and 3,000 employees, with $600 million the goal for 2001. With such explosive growth, Trend is confident that it will reach its billion-dollar goal in a relatively short time.

Trend's success is based in large part on having targeted a niche market that is global in scope and growing at a phenomenal rate — corporate outsourcing of manufacturing. The company has concentrated on a focused customer base that includes some of the biggest and most important manufacturers in the computer and electronics industries: Sun Microsystems, Silicon Graphics, Hewlett-Packard, Cisco Systems, Compaq Computer Corp., Dell Computer, Hughes Electronics and Nortel Networks. For these international giants, Trend is able to build multiple sets of tools to fabricate new products and speed them to market in the shortest time possible. Trend has the tooling capabilities for crafting parts in a wide range of sizes, from as small as a fingernail to as large as a computer chassis, making them quickly and delivering them fast.

Even with its hectic production schedule, Trend finds time to work with the communities in which its facilities are located. In Mexico, Trend organized to improve pavements, street lighting and police surveillance. To provide medical care for New Mexico's children, Trend makes donations to its Santa Fe Credit Union. Among other community services which benefit from Trend's support are police and sheriffs associations, athletic programs for children and teens, chambers of commerce and the YMCA. Charitable causes include the MS Walk, Toys for Tots, Juvenile Diabetes Foundation and local hospice programs. The company invests in the Society of the Plastics Industry Plastics Education Program to promote the understanding of the role that plastics play in our environment and lifestyles.

The future is promising for Trend Technologies. The company has experienced tremendous growth in just a few years, due to its commitment to giving its customers the absolute best in service and quality. While Trend has been awarded a long list of performance awards from its many satisfied customers, its success is the result of building partnerships that establish longtime loyalty, supplying the tools that create cutting-edge products, delivering them quickly and pursuing the vision to think globally and support locally in order to serve an ever changing world market.

The complex tooling operation at Trend Technologies' Tool Tech division

Top Line Electronics, Corp.

FOR PAUL KHAUV, FOUNDER OF TOP LINE ELECTRONICS, CORP., THE DECISION TO START HIS OWN electronics manufacturing company was a relatively easy one. After 10 years of work in the electronics industry in San Jose, he realized he had acquired plenty of business acumen and industry knowledge. Why not use his talents to run his own business?

Khauv and six others did their market analysis homework. They observed that the 1980s ebb and flow of the high-technology industries, which included computers and electronics, compelled many companies to downsize their work force, a key factor in keeping their total costs down. Many large established companies reviewed long-range strategies and began to outsource manufacturing services. Reading the trends of the past, and grasping what opportunities for manufacturing services lay in the future, Khauv started his own manufacturing company in 1989. In 1995, Top Line was certified by the International Organization for Standardization and in 1999 was certified by the British Approvals Board for Telecommunications (BABT). At the 10-year mark, Top Line employed 400 people and was ranked in the top 50 electronic manufacturing service providers in North America.

As a result of its partnerships, Top Line evolved into a turnkey manufacturer, a significant milestone for the company. Initially customers brought their product designs to Top Line for manufacturing only. However, as companies increasingly looked to outsource, and in order to adapt to clients' changing needs, Top Line expanded its services and its value in the competitive market.

As a turnkey manufacturer, Top Line is involved from a client's product concept to marketable unit. The company works with clients to develop the prototype, and is responsible for component procurement, manufacture of the product (board level assembly, chassis assembly, box build), product testing and shipping the line of goods worldwide. In its role as a turnkey solution service, Top Line wants to send a quality product to market quickly and cost-effectively. The time-to-market is crucial in maintaining an edge on the client's competition, enjoying a better market share and, of course, satisfying each client.

Needing more space, Top Line moved its corporate headquarters and manufacturing plant to a new San Jose location in the fall of 1998. In 1999, when additional manufacturing space was acquired — more than doubling its manufacturing area to 100,000 square feet — Top Line Electronics expanded into high-volume mechanical assembly (box build). Added Fuji SMT lines increased its manufacturing capability to handle a wider array of technologies such as Ball Grid Array (BGA) or fine pitch components found in palm-sized electronic equipment. Top Line's X-ray inspection and equipment testing capabilities were also augmented.

While facilities and services at Top Line have grown exponentially during its first decade, customer service remains the company's top priority. Khauv, who now serves as senior vice president of operations, and Daniel Chen, president and CEO, believe

SMT Lines —
Automated lines for Surface Mount Technology

that clients are business partners. Long-term relationships are desirable for both the customer and company, and to nurture that relationship, each account is assigned a specific customer service specialist who knows the detailed history of that company. If a client needs assistance on a special project, it is not uncommon for Top Line to send representatives to assist that client on site. Employee commitment is evident — more than once they have chosen to work over holidays and weekends to ensure that company operations wouldn't be interrupted and clients' schedules would be met.

Wave Solder

The company has seen the tangible rewards of this business partner strategy — business for Top Line doubled within an eight-month period from its existing customer base and two of its clients have gone public. With annual sales hitting $53.6 million in July of 1999, Top Line embraces its growth.

Regardless of how the company grows, Khauv and Chen agree that during this time of transition, the greatest challenge for the company is to maintain the flexibility that is available in a small to medium-sized company. They have watched similar-sized companies lose the ability to work closely, quickly and smoothly with their customers on the way to becoming a large entity. Top Line executives prefer to establish a built-in pliability by replicating the overall company structure and organization into each department.

Top Line believes in taking care of its employees, whose contributions have garnered more than two dozen industry awards in 10 years. The company offers full benefits and competitive salaries and creates a work environment that is comfortable and aesthetically pleasing — orchids and fountains decorate the offices, with a relaxing outdoor patio area provided for breaks. Employees come from all over the world, and it is not unusual for two or more members of the same family to be working on site. Top Line executives like family members working together because they can carpool, have lunch together or visit with each other during the day. An open-door policy keeps the top administrators approachable to staff, and staff members recount numerous times when the president and executives have shown genuine concern for their employees' personal lives. The attitude of the company's leaders has been effective — employee turnover is well below industry average.

Khauv and Chen look forward to personal time to interact in the community and for now, Top Line contributes to the Asian Americans for Community Involvement (AACI). AACI is a nonprofit advocacy, health and human services organization concerned with the welfare of Asian Pacific Americans. The organization provides bilingual services and programs for adults and children. Most of AACI's clients are Asian and Southeast Asian refugees and Khauv is proud that Top Line has helped provide the opportunity to learn job and language skills.

The future for their company, Khauv and Chen predict, is that Top Line will play a significant role as a total solution contract manufacturing company of the growing electronics industry in Silicon Valley. As Top Line keeps pace with the widening electronics industry by developing new business partnerships and services, it will continue to provide established clients with manufacturing expertise to enhance current products.

Bliss Industries, Inc.

Bliss offers 250 carts and racks to meet any assembly line need.

The new Flexbench by Bliss Industries

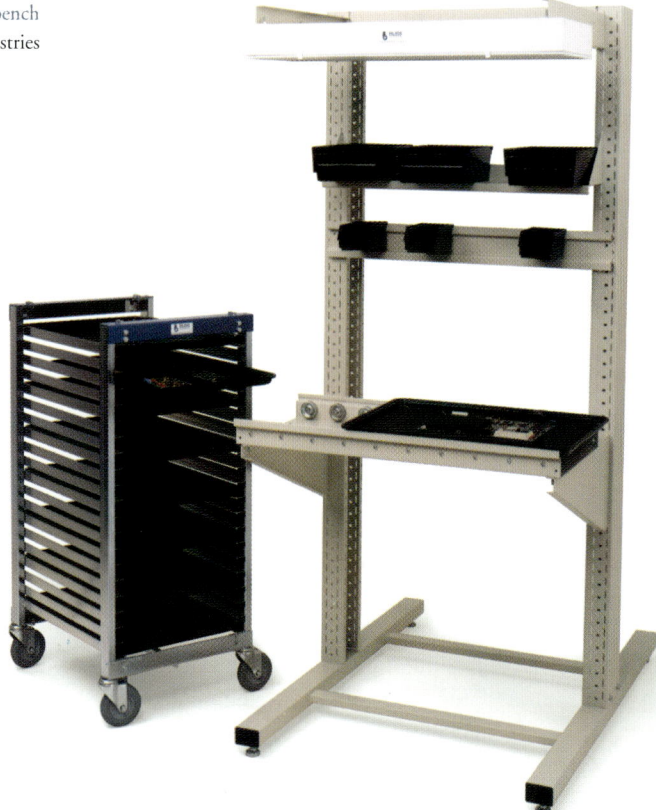

BLISS INDUSTRIES, INC., FOUNDED BY KEN BLISS ON APRIL 1, 1980, EMBARKED ON SEVERAL BUSINESS VENTURES, including precision sheet metal fabrication, before finding its role as a leading supplier of PC board assembly line solutions. In 1984 Apple Computer, then in its infancy, asked Bliss Industries to create a cart that would aid workers in printed circuit board assembly. Bliss' resulting product pleased the burgeoning computer company, and from that small beginning Bliss has evolved into its role as the preferred supplier for carts and racks that support PC board assembly.

Bliss continued creating carts and racks for Apple while expanding its product line, and in 1991 the company developed a tray cart, similar to a bakery cart, that literally reinvented the way PC board assembly is conducted. Customers raved about the new design that not only eliminated damage to PC boards caused by mishandling, but also protected against electrostatic discharge damage.

In 1992 Bliss took a quantum leap of faith in its ability to tailor products to meet companies' unique requests and launched a nationwide marketing program. Bliss distributed brochures on its product line at trade shows and gained many new customers. For three consecutive years *San Jose Business Journal* named Bliss Industries one of the top 100 fastest-growing privately held companies in the San Jose/San Francisco area.

Today, with over 250 products, Bliss is the only supplier of carts and racks to support every step of PC board assembly — from the receiving door, where the unassembled components start, to the shipping door, where the assembled boards go to market. Over 1 million times a day workers worldwide use Bliss carts while assembling PC boards that are found in new home appliances, pagers, cell phones, computers and Internet routers.

Bliss' key product lines are PC board assembly tray carts, stencil frame handling carts, feeder handling and storage carts, component reel storage, solder pallet handling carts, dross and fume containment carts, chip tube handling carts and PC board magazine handling carts.

In 2000, celebrating its 20th year of operation, Bliss introduced a completely versatile addition to the cart product line, "Flexbench." Described as "the ultimate techbench," Flexbench can be used to create a workstation of any size for any process in the electronics industry.

All Bliss products have been created in response to customers' specific requests. But Bliss is more than just a supplier of carts for electronic assembly. Based on its engineering experience working with and developing applications for hundreds of customers, Bliss offers invaluable expertise to businesses that want to operate more efficiently and profitably. Its zeal in converting lost production time into increased plant capacity and profits won Bliss Industries the coveted SMT Vision Award in 1996, and its ability to respond quickly and dynamically to customer demands for assembly line solutions ensures Bliss a continuing presence in the growing computer industry.

Palmer College of Chiropractic West

JUST AS THE 1980S CONTINUED SANTA CLARA VALLEY'S EVOLUTIONARY SHIFT FROM AN AREA RENOWN more for its industry than agriculture, this decade also triggered a significant shift in societal attitudes about health care. And just as many of the companies instrumental in spurring this high-tech shift were based in the Silicon Valley, another area-based institution was playing a prominent hands-on role in the "alternative" health care boom, in large part due to its high-touch approach to patient care.

Palmer College of Chiropractic West (PCCW) was founded on October 17, 1980, on the former Peterson Junior High School campus in Sunnyvale. Palmer West shares its name, philosophy and tradition with the profession's oldest chiropractic college, Palmer College of

(Far right) Resources in the college's library include a CD-ROM-based computer with Medline and the "Medical Interactive" computer system with case studies and more than 40,000 images from the American College of Radiology Learning Library.

(Right) Situated in the heart of the Silicon Valley, the Palmer College of Chiropractic West campus projects an image that reflects the high-tech design of its innovative and cutting-edge neighboring companies.

Chiropractic, in Davenport, Iowa. In May of 1991 the academic mission of the two Palmer Colleges was strengthened by the unification of the two institutions to form the Palmer Chiropractic University System.

Palmer West is accredited to award the doctor of chiropractic (D.C.) degree by the Commission on Accreditation of the Council on Chiropractic Education. Palmer West's patient-focused Clinical Practice Curriculum is a 13-quarter (3.25 calendar years) program that includes extensive and intensive study of the basic and clinical sciences. In addition to assimilating this knowledge and applying patient-care skills in a pregradate internship, Palmer West's curriculum also prepares its students for the "real world" by providing pregraduate field training in the office of a practicing doctor of chiropractic.

As chiropractic has grown, both in its number of patients and practitioners (the profession now represents the third-largest group of primary-care providers in the United States), so too has Palmer West's student body. The college's 1981 enrollment was 148 students. In 1989, increasing enrollment prompted the college to expand and occupy a second campus on Pomeroy Avenue in Santa Clara. In 1993 Palmer West consolidated its two-campus format and moved into its current campus on Tasman Drive in San Jose. By January of 1998, student enrollment at Palmer West had increased to more than 700 students.

Just as the chiropractic profession has matured from an "alternative" form of health care to its contemporary status as a "complementary" discipline (recent studies by

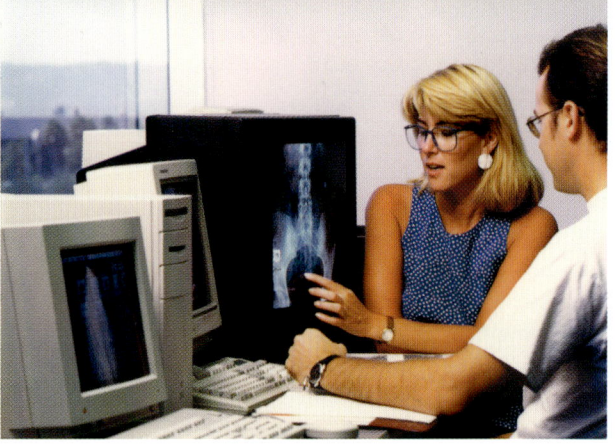

RAND Corporation and the Federal Agency for Health Policy and Research have shown chiropractic's clinical and cost effectiveness), the demographics of those pursuing their doctor of chiropractic degree at Palmer West has also seen a significant change during the past 20 years. In 1983 the average age of an entering Palmer West male student was 29. Today that figure is closer to 25. While more women are entering the profession, men continue to represent nearly 75 percent of the student body.

While the majority of Palmer West students originate from the western United States, international enrollment accounts for nearly 20 percent of the PCCW student body (compared to 3 percent in 1983). More than two dozen flags, representing countries from which Palmer West students have traveled to pursue their chiropractic studies, hang in the main lobby of the college campus. Now approximately 3,000 Palmer West graduates are impacting the quality of health care and the quality of life for patients all over the world.

In addition to a campus-based clinic, which draws from the neighboring industrial park companies, Palmer West also maintains outpatient clinics in downtown Santa Clara and downtown San Jose. As primary-care, portal-of-entry providers, doctors of chiropractic focus their attention on musculo-skeletal conditions. While the spinal adjustment represents the doctor of chiropractic's primary patient-care skill and service, the Palmer West program prepares its graduates for the duties and responsibilities that come with primary-care provider status. The college curriculum includes extensive study of the basic and clinical sciences, including diagnostic and radiographic procedures.

In 1998 Palmer West's three outpatient clinics recorded more than 35,000 patient visits. As the hub of its community outreach program, Palmer West's clinic in downtown San Jose is symbolic of the college's commitment to the community it has served during the past two decades. More than 4,000 patients have visited PCCW's outreach clinic since its opening in 1992. The college annually contributes more than $100,000 in complimentary patient-care services for homeless and financially disadvantaged families throughout the South Bay Area.

For many, Palmer West's Outreach Clinic is their only avenue of health care. Said one recent patient, "I consider the Palmer outreach clinic an act of caring, a humanitarian effort... Without this clinic, where would the many in need go? Palmer has opened its arms to the homeless, or anyone in need. You can't top this standard of health care."

Other examples of Palmer West's commitment to the community are its satellite clinics based at CityTeam Ministries and Salvation Army adult rehabilitation centers in San Jose. "This is another way Palmer West is giving back and staying in touch with the community," said Dr. Andre KnustGraichen, primary attending physician of the Palmer West Outreach Clinic. "By taking a 'hands-on' role in the recovery process, we're not only saving money for the community, we're also assisting the residents through their recovery process and helping to guide them toward a constructive rather than a destructive way of life."

In addition to its outreach and satellite clinics, Palmer West also serves the South Bay through other community programs. One of the college's truly "special" programs is that which it maintains with Special Olympics of Santa Clara Valley. Palmer West helps Special Olympic athletes meet their physical examination prerequisite for event participation by providing complimentary physical exams. Palmer West's Sports Council, comprised of upper-quarter students who have demonstrated special skills in the care of common sports-related injuries, provides on-field care, under the direct supervision of Palmer West clinical faculty, for athletes and volunteers at Special Olympic events.

"Having a resource like Palmer West for our athletes is a dream; it has helped to boost the number of athletes who are now able to complete in our events," said Kara Capaldo, Special Olympics training coordinator.

With a new millennium underway, and chiropractic beginning its second century of service, Palmer West stands as both a link to the valley's past and an example of another Silicon Valley institution that is innovatively "adjusting" the future and providing services that are making a worldwide impact.

Palmer College of Chiropractic West, an accredited institution founded in 1980, conducts a 13-quarter (3.25 years) program for approximately 800 students on its main campus on E. Tasman Drive in San Jose.

Palmer West has a strong commitment to serve the South Bay community. For the past seven years the college's Community Chiropractic Clinic in downtown San Jose has provided complimentary care for thousands of financially or otherwise disadvantaged patients, many of whom have no other form of health care assistance.

San José State University

DURING THE 1800S, INSTITUTIONS OF HIGHER LEARNING IN AMERICA WERE DESIGNED TO INSTRUCT AN upper-class gentry in classical literature and languages, graduating a "Christian scholar and gentleman" who typically had little need to work for a living. But by the late 1850s and early 1860s, the country's economy shifted from an agrarian to a technological base. The landed aristocracy of America all but disappeared and was replaced by a burgeoning population of men and women who earned their own livings, raised and educated their own children, enjoyed ample leisure time and took an active part in local, state and federal government.

Classic Tower Hall is the heart of the SJSU campus.

As the country moved to democratize higher education, the need for trained teachers escalated. The existing secondary schools provided some instruction in the techniques of teaching. But their lack of success in adequately staffing the public schools led to the establishment of community-supported institutions focused on training public school teachers. Patterned after the European *École Normale*, literally "model school" but Americanized as "normal school," the Minns' Evening Normal School opened in 1857 in San Francisco. That was the birth of what came to be known as San José State University.

On May 2, 1862, the California State Normal School was created by legislative act, shifting the responsibility of training public school teachers to the government. Coursework included: elocutionary exercises, reading, composition, arithmetic (mental and written), map drawing, human physiology and the science and art of teaching. Hopes for a rush of enrollment in the new school, opened that July, were dashed when only six students appeared.

But when State Superintendent John Swett announced in June 1867 that, for the first time in California history, every public school would be entirely free for every child (property taxes would be used to raise the funds), demand for professional teachers skyrocketed. Enrollment in the California State Normal School then grew rapidly.

The school's move to San José was triggered on October 21, 1868, when a severe earthquake struck San Francisco, causing considerable destruction throughout the city. Although damage to the normal school itself was slight, the State Board of Education decided to move the school to a location thought to be less prone to earthquakes.

The new location had to be determined by the legislature. San José, Santa Clara, Vallejo, Stockton, Martinez, Oakland and other cities made aggressive bids. It took five votes by the legislature, but in March of 1870 the Senate and Assembly voted in favor of San José, a

A culturally diverse student population of over 67,000 provides a unique educational experience.

thriving town of 9,089. Washington Square in the heart of town was selected as the site and the cornerstone was placed on October 20, 1870. The school formally reopened on June 14, 1871, in temporary quarters — space provided at the Santa Clara Street School by the city Board of Education.

Finally, on June 7, 1872, the California State Normal School moved into its new building on Washington Square to begin life as a self-sustaining educational institution. The school grew steadily, adding students, faculty, new coursework and facilities and before long it was ready to expand geographically. In March of 1880 a Branch State Normal School was approved for the city of Los Angeles. As fate would have it, the Los Angeles State Normal School became the southern branch of the University of California in 1919 and then the University of California at Los Angeles (UCLA) in 1927.

At a turning point in June of 1921, the Board of Trustees, which had managed the school for 59 years, relinquished its responsibilities to the State Department of Education and announced that the school would be known as San José State Teachers College. By the 1929-1930 academic year, more than 3,000 students attended the college. And on September 22, 1930, the first program leading to a college degree for police officers was started at San José; it was the first such program in the world.

The state legislature authorized changing the names of teachers colleges to state colleges in June 1935, so the moniker San José State College was born. Though the war years kept enrollment low, by 1948 more than 7,000 students were enrolled. The following year, the college received permission to offer a B.S. degree in engineering (a million-dollar engineering building would be dedicated in February 1954). In June 1950, 11 graduate students earned the first M.A. degrees from San José State. In 1955 master's degrees were awarded in engineering. From 1948-1958 enrollment doubled.

The school adopted its current name on January 1, 1974. Governor Ronald Reagan signed a measure changing the name to San José State University (SJSU).

Today, San José State University is the largest educational institution in the Silicon Valley. Eight colleges offer 134 degree and 29 certificate programs to a culturally diverse student population of 67,000 (including continuing education students). The University has built industry partnerships resulting in custom MBA programs delivered at corporate sites, business incubators and a Sun Microsystems Unix Certificate program. And still true to its original mission, SJSU's College of Education credentials about 650 teachers each year.

In a break from traditional thinking, SJSU has formed a partnership with the city of San José to build a new $175 million library. The eight-story library facility, designed, operated and funded jointly by the university and the city of San José, is scheduled to open in 2003. The first such collaboration between a major university and a large city, it will be home to both the main branch of the San José Public Library and the entire university collection (currently split between two buildings).

SJSU students will have twice the study space previously available and every seat will be wired for connection to the Internet. With room for 25 years of collection growth, the library stands not just as the culmination of a long history of progress, but as the next step in San José State University's commitment to delivering innovation and excellence in higher education.

Valley Transportation Authority

WINE, SPEECHES, A BRIGHT, SUNNY DAY: NOVEMBER 1, 1868. SAMUEL ADDISON BISHOP, THE FATHER OF SAN JOSE'S local transit system, was hosting a few friends and notables for a ride on the newly launched San Jose and Santa Clara Railroad. In a horse-drawn railcar, the three-mile trip lasted about 45 minutes. Three days later, regular service between the two young communities would be established and interurban public transportation in the Santa Clara Valley would be born.

Like so many of the country's great industrial initiatives, public transportation would suffer fits and starts, business rivalries, political battles, good and bad ideas, second-guessing and regrets. A few privately owned, for-profit horse-car companies evolved into quite a few city and suburban electric tram lines, which were consolidated by the Southern Pacific Railroad in March 1912. After pouring millions of dollars into the system, however, Southern Pacific ultimately was undone by the Great Depression, closing its electric tram routes for good in April 1938.

In the period just before and right after World War II, railroads in most U.S. cities were scrapped because upkeep and upgrade costs were too high. Bus lines were created to serve urban commuters, but the economics of small, individual, unlinked bus operations eventually led these bus lines to the brink of failure as well. By 1969, local lines around the Santa Clara Valley struggled to continue operations and were heavily subsidized. The idea of creating a new countywide agency to buy out the local operators and run the buses began to gain popular support.

The California Legislature passed the Santa Clara County Transit District Act in 1969, sanctioning the takeover of three private bus lines and expansion from 60 to 325 buses by 1985. However, a ballot measure needed to create the new agency and authorize a half-cent sales tax was defeated. In 1970 a similar measure also was defeated. Subsidies to operate buses continued to be paid out of property taxes.

Enter the U.S. government. In late 1971, Congress began to support the use of federal funds to prime the development of urban transit systems. In California, the Transportation Development Act was passed to apply the state sales tax to gasoline and channel those funds to public transit initiatives.

On June 6, 1972, voters approved the new Santa Clara County Transit District. The margin of victory was better than 2-to-1 and County Transit made its debut.

Building and operating a mass transit system is an imperfect science, especially when the populace being served is growing rapidly and becoming more widespread. County Transit tried new programs and, in the main, enjoyed great success, although there was one notable failure.

Car 124 originally ran in San Jose from 1912 to 1934, when it was sold (along with Car 73) for use as housing on Old Almaden Road. It was reconstructed and returned to service by the San Jose Trolley Corporation on November 18, 1988.

The Dial-A-Ride program was too successful for its own good. Launched in November 1974, it offered one-call, computer-dispatched, door-to-door service to bring people from outlying areas of the county to main bus routes. Daily demand was expected to be about 10,000. On opening day 80,000 calls came in, and 80,000 calls per day became the norm. After a few months, analysis showed that the per-passenger Dial-A-Ride cost per mile was more than eight times the per-mile cost on main arterial routes. Emergency funding and adjustments notwithstanding, the great idea that had been implemented and was providing a popular service had two major unsolvable problems: it was oversubscribed and too expensive to maintain. In mid-May 1975, except for a few routes in the far southern part of the county, Dial-A-Ride was discontinued.

Counterbalancing the demise of Dial-A-Ride was voter approval of a permanent half-cent sales tax in 1976 for transit operations. This dedicated local revenue source provided a firm financial basis for building a comprehensive transit system.

The next era of public transit for Santa Clara County began with the introduction of light rail. A broad consensus of city leaders, regional agencies, voters and Silicon Valley high-tech industries supported the Guadalupe light rail project. Support also came from Sacramento and, later, the federal government.

After some six-and-a-half years of planning and consensus building, ground was broken on March 12, 1984. By June 1988, light rail service from Santa Clara/North San Jose to San Jose's revitalized downtown was opened for riders, and in August 1990 the first section of the light rail line to serve a residential neighborhood opened.

On April 21, 1991, seven years after the first groundbreaking ceremony, the entire 21-mile light rail line opened for service. More than 15,000 people attended the ceremony. Santa Clara County had a light rail line that ran from suburban neighborhoods through downtown San Jose and north to the rapidly growing high-tech areas of Silicon Valley.

With passenger numbers growing and new light rail construction under way, 1995 was a banner year for County Transit. On January 1, the Transit District merged with the Santa Clara County Congestion Management Agency (CMA). Consolidated under a single governing board and separate from county government, this new agency became responsible for multimodal (more than one type — buses, light rail, trains, etc.) countywide transportation planning, including land-use and transit operations. The new board, a consortium of 15 cities along with county representatives, adopted a broader, more regional approach to transportation operations and planning.

In January 1996, the agency adopted a new name: Santa Clara Valley Transportation Authority, and that December the VTA logo was adopted.

In December 1999, VTA made its latest contribution to providing a comprehensive transportation system in Santa Clara county: the 7.6-mile Tasman West light rail extension from North First Street in San Jose to Downtown Mountain View opened for regular service. With the addition of Tasman West, VTA's operations comprised a 525-bus fleet, rail shuttle and paratransit services, and regional rail (Caltrain, ACE) and bus (Dumbarton, Highway 17) services. In addition, VTA embarked on implementation of the 1996 Measure B Transportation Improvement Program, a comprehensive program in partnership with the county of Santa Clara, including: three more light rail extensions, a commuter rail connection to BART in Alameda County, several highway expansions, bikeways and local street repairs.

Reflecting the shift from transit provider to countywide transportation planner (including responsibility for buses, rail systems, cars, cyclists and pedestrians), VTA also began developing the Valley Transportation Plan 2020 (VTP 2020) in 1999.

The unprecedented growth of the Silicon Valley, along with the increased demand for travel "24 X 7" (24 hours a day, 7 days a week) and the shrinking availability of land, puts intense strain on the entire transportation system. Resources have to be balanced between public transit and roadway projects. New technologies have to be predicted, proven and implemented, and land use and transportation developed compatibly.

VTP 2020 provides a comprehensive 20-year plan for financing, system management, transit services and land use. The plan is dynamic; it will be updated every two years. The preparation and planning that went into VTP 2020 are the results of more than 100 years of experience by VTA and its varied forebears. All have gained considerable experience in contending with one unyielding constant: the burgeoning growth of the Santa Clara Valley.

The Santa Clara Valley Transportation Authority (VTA) is an independent special district responsible for bus, light rail and paratransit operations; congestion management; specific highway improvement projects; and countywide transportation planning. As such, VTA is both a transit provider and multimodal transportation planning organization involved with transit, highway and roadways, bikeways and pedestrian facilities.

History San José

HISTORY SAN JOSÉ INVOLVES DIVERSE AUDIENCES IN EXPLORING THE VARIETIES OF HUMAN EXPERIENCE THAT CONTRIBUTE TO THE CONTINUING HISTORY OF SAN JOSÉ AND THE SANTA CLARA VALLEY.

History Park is a charming 14-acre site complete with paved streets and running trolleys located in San José's Kelley Park.

VALLEY OF THE OAKS, SANTA CLARA VALLEY, VALLEY OF HEARTS DELIGHT, OR SILICON VALLEY — WHATEVER it is called, there are few places in the country that have as long, deep and multitextured histories as the fertile plain at the south end of San Francisco Bay. At the same time, there is no place as busy inventing the future for the entire world. San José is the oldest civil settlement in California and the state's first capital. It became capital of Santa Clara Valley's rich fruit orchard and processing operations. Today it is capital of the world's computer and Internet industries. History San José is the dynamic organization bridging the unparalleled heritage with the boundless future of this remarkable valley.

The history of San José's historical museums spans one-third of California's years in the American Union. Upon the sesquicentennial of the California Gold Rush in 1949, the city of San José erected a replica of California's first State House on the downtown plaza. Twenty years later, in 1971, the city opened the San José Historical Museum in the south end of Kelley Park. Over the past three decades, 28 historical buildings have been relocated or reconstructed on the museum's 14 developed acres, now called History Park.

In 1971 the San José Historical Museum Association was founded to raise funds in support of buildings, programs and activities at the Kelley Park museum site. In 1993 the association contracted with the city to manage the Peralta Adobe and Fallon House museums on St. John Street, at the north end of San José's developing downtown. Today, History Park, the Fallon House and the Peralta Adobe all are managed by History San José.

The 1797 Peralta Adobe, one of the oldest residences in California, and the 1850s stately Italianate Fallon House across the street welcome visitors through their doors. Both are within walking distance of the downtown hospitality and entertainment center. The Peralta Adobe is the first historic site on the Juan Bautista de Anza National Historic Trail, the only Latino trail included in the National Parks' National Historic Trails program.

Visitors enjoy riding a trolley, attending a one-room school and experiencing one of the many festivals in History Park, about a mile south of downtown. History Park is dedicated to understanding and sharing the variety of cultural experiences and communities that have come together to define this dynamic and diverse region. The park includes a historic Chinese Temple and a Portuguese Imperio and hosts many special ethnic and cultural festivals throughout the year. And, most days History Park is free.

Beautifully restored, the Peralta Adobe, built in 1797, is the city of San José's oldest home.

History San José also showcases the area's history through its Web site. Whether virtual collages created in innovative adult literacy workshops, online exhibitions or women's history pages produced by high school students, the Web pages are innovative and entertaining and take history to new people in new ways.

This is History San José — Silicon Valley from A to Z — taking the past into the future and redefining the historical museum in America for the next millennium.

SAND DREAMS & SILICON ORCHARDS

San Jose Water Company

FOR NEARLY 135 YEARS, SAN JOSE WATER COMPANY HAS SERVED RESIDENTS OF THE SANTA CLARA VALLEY. Today, the company bears the distinction of being California's oldest investor-owned water utility and San Jose's second-oldest continuously operating business. Though it began with only 400 customers, the company now has 216,000 service connections and roughly 1 million customers in its 134-square-mile service area. Regulated by the California Public Utilities Commission (CPUC), the privately owned public utility provides water to much of the valley — most of San Jose and Cupertino, and all of Campbell, Los Gatos, Saratoga and Monte Sereno.

Artesian wells were the building blocks of San Jose Water Company. The first artesian well in San Jose, dug in 1854, brought forth a stream of water in such force as "almost sufficient to run a saw mill." In 1864 Donald McKenzie, a shrewd Scotsman and foundry owner, saw a tremendous opportunity to capitalize on the rich natural resource that lay directly beneath San Jose's city streets. He gained permission to build two wooden tanks at the San Jose Foundry and he filled them with well water. With two other men, John Bonner and Anthony Chabot, McKenzie founded San Jose Water Company with $100,000 in capital. The company was granted exclusive privileges — which included the franchise to run pipes under the streets — by the city of San Jose and the town of Santa Clara to serve their residents with "good and pure water." Articles of incorporation were signed on November 21, 1866.

Within three years the capital was increased threefold "to meet the extensive development planned by the company." Realizing that artesian wells would not supply a sufficient amount of water, San Jose Water Company authorized the construction of a flume to carry runoff water from the Santa Cruz Mountains into a local reservoir.

San Jose Water Company has grown with the community it serves through mergers and the purchase of additional water rights and watershed expansion areas. Today, San Jose Water Company obtains 45 percent of its water from the Santa Clara Valley Water District, which imports water from the Central Valley and then treats it. Another 45 percent is comprised of ground water pumped from the local aquifer and then chlorinated by San Jose Water Company's pumping stations. The remaining 10 percent comes from mountain runoff water chlorinated by San Jose Water Company's two treatment plants in Saratoga and Los Gatos.

Under J.W. Weinhardt, president from 1974 to 1995, the company advanced technologically to provide better customer service. The billing system and operation of both the pumping facilities and the distribution system are fully computerized. Additionally, meter readers input their findings into hand-held computers that are downloaded directly into the billing system.

In October 1999, with Richard Roth as president, San Jose Water Company entered into a merger agreement to be purchased by American Water Works, the largest water company in the United States. The acquisition is pending approval of the CPUC. Regardless of ownership, the company will continue its 135-year-old legacy of providing "good and pure water" without serious interruption.

The original tanks of the San Jose Water Company, c. 1866

The landmark San Jose Water Company office building, c. 1934

Symmetry Corporation

IN DEVELOPING A GLOBAL BUSINESS THAT HAS ACHIEVED MORE THAN $100 MILLION IN SALES IN LESS than five years, Rudy Revak, president and founder of Symmetry Corporation, has proved true the aphorism "Do what you love; the money will come."

Revak was born in Altrusried, Germany, the son of Hungarian and Romanian refugees who fled the Communist takeover following the end of World War II. Revak's grandparents came to the United States and then sponsored Revak's parents. After serving in the U.S. Army in Vietnam, Revak went to school under the G.I. bill. Then, while also holding a full-time job, he attended an invitational meeting of a soap company that sold its products via direct selling. Revak began part-time work selling soap for the company. Soon he was doing well enough to quit his other job. He became an international vice president and traveled to Asia, Europe, South America and throughout North America to build the company's business. Although successful, Revak wanted to come home.

Meanwhile, in 1975, after trying different nutritional supplements, Revak found that he felt better and had much more energy than before. He convinced his friends of the benefits of the supplements and asked himself, "Why not sell them?" At this time, the company he worked for owned a jewelry business in Los Gatos, California, that was losing money. Revak was sent to take charge of it. Over a three-year period, Revak evolved the business's focus from jewelry to skin care products to nutritional supplements, building sales from nothing to $60 million in a year, through person-to-person marketing.

In March 1995, Revak left the thriving company to pursue his entrepreneurial dreams. On May 15, 1995, he opened his own nutritional supplement business in Milpitas. Symmetry Corporation was founded on Revak's fundamental beliefs in integrity, empowerment, equality and love. Symmetry, with 180 employees worldwide, now does business in 10 countries — the United States, Trinidad, Jamaica, Barbados, Saint Lucia, Bahamas, Canada, Philippines, Mexico and Hong Kong. Symmetry adds England in 2000 with future expansion planned for Europe, South America, Asia and Australia.

Symmetry's products are manufactured using the most advanced techniques under stringent guidelines to ensure that consumers receive the freshest, purest and highest-quality product possible. In addition to a comprehensive line of nutritional supplements, the company sells skin care products and water enhancement systems. Symmetry has put together a team of experienced executives and tens of thousands of distributors who support Symmetry's mission to enrich people's lives for a better tomorrow physically, mentally, personally and financially.

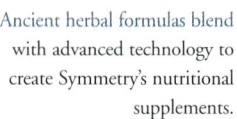

Rudy Revak, president and founder of Symmetry Corporation

Ancient herbal formulas blend with advanced technology to create Symmetry's nutritional supplements.

SAND DREAMS & SILICON ORCHARDS

University of California, Santa Cruz

ADDING TO ITS CALIFORNIA REDWOODS, 2,000 ACRES OF MEADOWS AND FORESTS, AND VISTAS OF THE sparkling Pacific Ocean, the University of California, Santa Cruz (UC Santa Cruz) is establishing a new regional center in the Silicon Valley. Hailed in 1998 as one of the "20 Best" public campuses in the country by *U.S. News and World Report*, UC Santa Cruz is now ready to expand its academic presence in San Jose.

UC Santa Cruz is not a newcomer to the Silicon Valley. Decades of collaboration between leading Silicon Valley partners and UC Santa Cruz students and faculty have provided this burgeoning area with a sturdy academic infrastructure. Each year more than 50,000 Silicon Valley professionals enroll in UC Extension classes in the greater San Jose area. These courses help professionals to keep up with the latest technological advancements and management skills, as well as offer creative self-expression.

Now, classes offered at the new San Jose site will lead students toward academic degrees. Students can anticipate development of both graduate and undergraduate programs. Partnerships with industry and other higher education institutions are in place or in the university's future plans, with the benefit of access to highly advanced facilities throughout San Jose. For the first time, local students enrolled in these programs can live close to their research and close to their classrooms.

Since 1965, when UC Santa Cruz was founded and subsequently acquired the Lick Observatory atop Mount Hamilton in eastern San Jose, the Silicon Valley has witnessed the many benefits of university programs. It was at this 100-year-old observatory that UCSC astronomy/astrophysics doctoral graduate Geoffrey Marcy was the first to discover planets in other solar systems. Other UC Santa Cruz research disciplines also have vital research links in San Jose, including the fields of biotechnology, electrical engineering, computer engineering, economics, environmental studies and politics.

Students and faculty in the Jack Baskin School of Engineering are pioneering the kinds of engineering that will be most needed in the 21st century. Included in the curriculum are computer engineering, computer science, electrical engineering, network engineering, information systems management and bioinformatics.
Photo by R.R. Jones

K-12 education also reaps rewards from the university's established programs. Partnerships with Eastside Union High School District and the San Jose Unified School District work to support curriculum development and increase resources for teachers. UC Santa Cruz-based scholarships, grants and partnerships with all of the region's community colleges have increased access to the university for many high school and community college students.

UC Santa Cruz leads a new University of California program called UC College Preparatory Initiative, also known as the "Virtual High School." This innovation makes advanced college preparatory courses available to students throughout California. So progressive is this cyberspace instruction that it may be foreshadowing aspects of future pedagogy.

As the population in San Jose swells to make it the third-largest city in the state of California, the addition of a University of California site seems like a natural progression that is long overdue. The success of UC Santa Cruz's partnership, extension and outreach programs in the Silicon Valley provide a rich history and a firm foundation upon which the university can build in the coming years.

UC Santa Cruz partnerships with several Silicon Valley school districts focus on enhancing teacher preparation and inspiring kids to prepare for a college education at an early age.
Photo by R.R. Jones

IDT (Integrated Device Technology)

ON ITS WAY TO BECOMING A $1 BILLION COMPANY AND CELEBRATING ITS 20TH ANNIVERSARY IN 2000, Integrated Device Technology (IDT) is a leading player in the communications semiconductor world. Based in Santa Clara, California, the heart of Silicon Valley, IDT is powering the products that are changing the way people communicate — at work and at home — with products aimed at key sectors of the communications markets. The company is a global semiconductor solutions provider to leading-edge communications companies — those that are driving innovation in the convergence of voice, data and wireless networks.

IDT's customer list reads like a Who's Who of the communications industry. The company's top 20 customers are among the world's leading communications suppliers — companies such as Cisco, Ericsson, Lucent, NEC, Nokia, and Nortel Networks, to name a few. IDT also works with many of the world's emerging leaders — those companies that are creating next-generation products that will include IDT's leading-edge solutions.

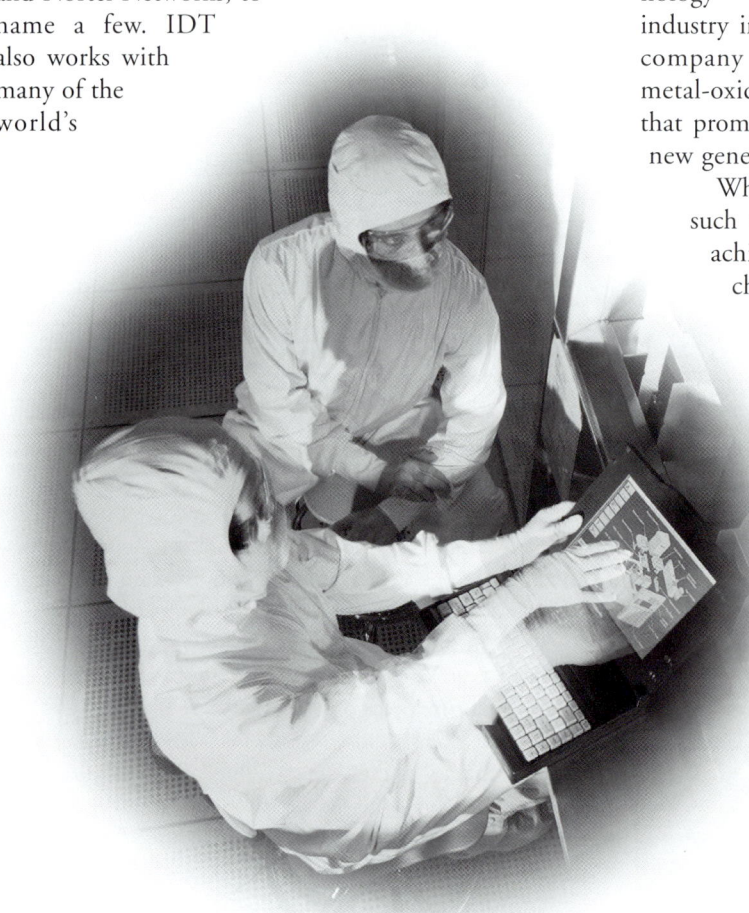

IDT's leading-edge manufacturing and process technologies produce a variety of high-performance products.

Why do these companies choose IDT? Its high-performance products help accelerate time to market and enhance the performance of systems in emerging data, voice and wireless networks. And IDT has a long history of powering leading-edge, high-performance systems in the communications and computing industries. For more than two decades, IDT has been pushing the technology envelope — and has strategically evolved its corporate vision to continually achieve successes.

IDT's FOUNDING STRATEGY

Right from the start, IDT set out to do something no one had done before with semiconductor process technology — the fundamental driver behind nearly all industry innovation at the time. Founded in 1980, the company revolutionized CMOS (complimentary metal-oxide silicon), a fast-evolving process technology that promised to dramatically reduce costs and enable new generations of more highly integrated ICs.

While CMOS was used extensively in products such as calculators and watches, it was an underachiever when it came to performance. IDT changed all of that by developing the first low-power, high-speed, high-performance CMOS device — an SRAM (static random access memory) with a 90-nanosecond access speed. And this technological feat, which literally put IDT on the map, was accomplished just a year after its founding.

With its high-speed CMOS technology, IDT was achieving significant performance breakthroughs. Through 1986, IDT was primarily a supplier of fast SRAMs to every major military prime contractor in the United States. The company also pioneered ICs for up-and-coming technology leaders. From Sun Microsystems' workstations to Silicon Graphics' servers and some of the first high-performance RAID controllers, IDT chips were powering the highest-performing systems at that time.

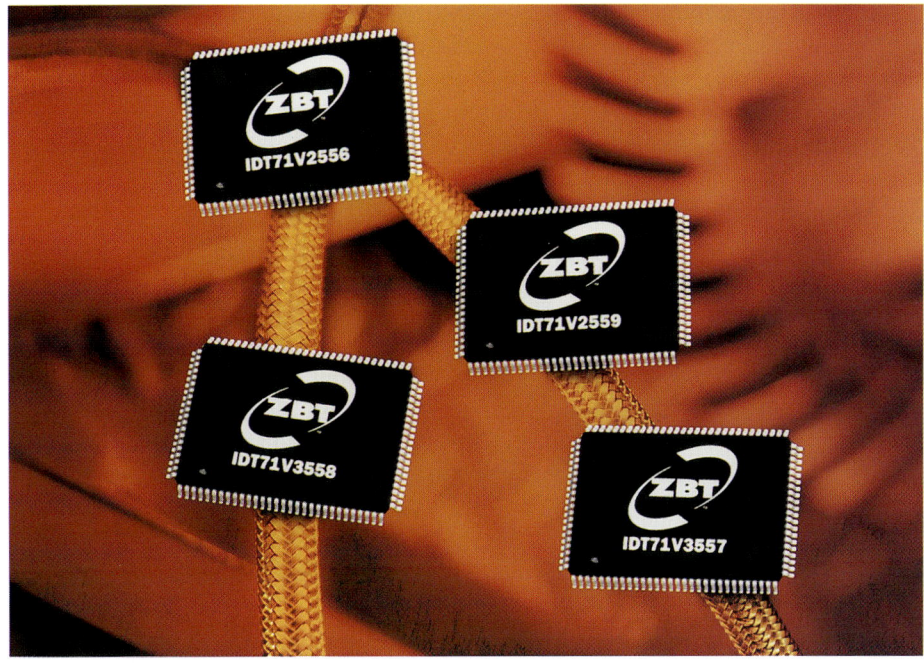

During this period, the demand for secondary cache to support powerful new generations of microprocessors exploded. IDT quickly became an industry-leading supplier of SRAMs for the booming PC (personal computer) cache business.

"In the early 1990s, by most measures, IDT was primarily considered a memory company because the lion's share of its products were high-performance SRAMs. That strategy led to tremendous growth for us," explained Jerry Taylor, IDT's president and chief executive officer.

In fiscal 1995, IDT achieved its first $100 million quarter, and in fiscal 1996 the company reported record revenues of $679 million. Life was good in the memory business.

IDT invented the Zero Bus Turnaround™ (ZBT®) architecture, the new standard for high-speed communications applications.

Without a doubt, the company was establishing a reputation for design leadership and process technology expertise. Yet the competition was figuring out the secret to fast CMOS technology and many serious players were entering IDT's game.

MEMORIES AND MORE

Recognizing that it had to broaden its offerings to stay competitive, IDT began offering other products such as logic devices, FIFOs and dual-port memories. In 1989, IDT strengthened its role as an industry pioneer by entering the RISC microprocessor market as one of the first licensees of the MIPS® architecture. At the same time, CMOS was becoming increasingly mainstream. As a result, CMOS costs were dropping and its uses were expanding well beyond niche applications. The ability to manufacture high-quality parts in commercial volume and price points became a key industry driver. IDT invested significant resources in its manufacturing organization — to complement wafer factories in California, IDT opened its Penang, Malaysia, assembly and test facility in 1988.

TRANSITIONING THE BUSINESS

In 1995, still in the midst of the SRAM boom and doing exceptionally well in the PC cache business, IDT began taking steps to transition into new areas in anticipation of changing market demands and to improve IDT's long-term viability in the market. Explains Taylor, "We

The SuperSync™ II family is part of IDT's successful communications memory offerings, providing the industry's highest data rates needed for leading-edge switches and routers.

recognized the importance of moving from a cost-driven, cyclical SRAM business to a more value-driven model. And we knew it would become more and more difficult to successfully compete in the commodity memory business."

In 1996 the semiconductor memory market fell dramatically with the price of SRAM plummeting nearly 90 percent by year-end 1996. And no company was immune from the impact. Cheap memory chips from overseas suppliers flooded the market. Almost overnight, dozens of companies went from having profitable memory businesses to reporting mounting losses. Margins didn't erode, they evaporated.

Despite those market dynamics, IDT used its SRAM success to reposition and strengthen its traditional businesses and to fund new product and market diversification efforts into new areas. The company saw that its future market growth would come from value-added products that offered high performance, differentiated capabilities and fast time to market. It concentrated on solving problems in the high-speed, high-bandwidth communications market — putting its engineers to work creating new types of communications memories and communication specific products, new approaches to architectural design, and new levels of integration.

IDT is powering the products that change the way the world communicates.

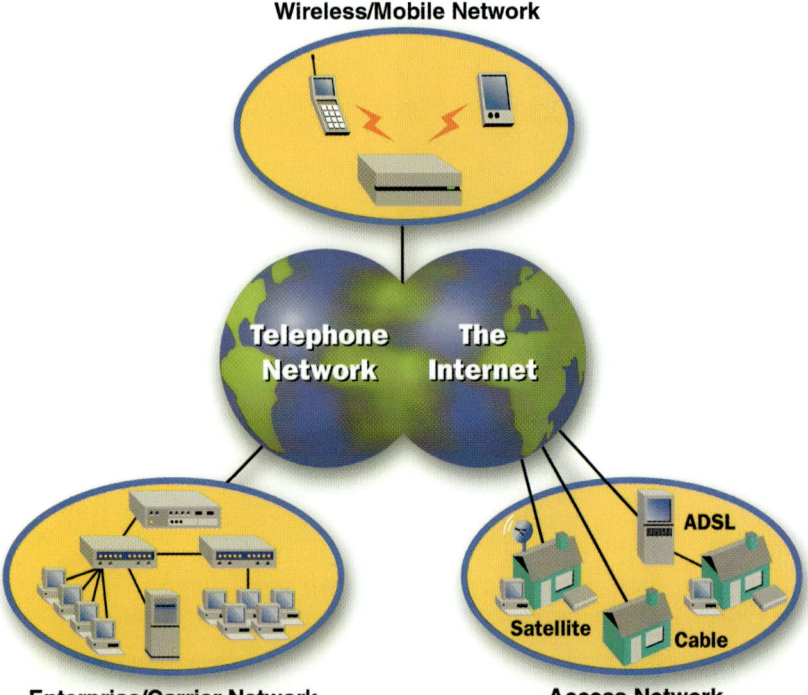

BANKING ON COMMUNICATIONS AND COMING BACK STRONG

In 1997 the company introduced the ZBT® (Zero Bus Turnaround™) architecture, which has become today's de facto standard in new designs with leading networking customers who require the ultimate in system bandwidth. In practical terms, IDT's ZBT architecture can actually double the effective performance of a high-speed switch. IDT now licenses this innovative architecture to partners Micron Technology and Motorola.

IDT also announced its highly integrated two-chip networking product, SwitchStar, a combination memory and controller used in low-cost ATM (asynchronous transfer mode) switches such as DSLAMs. This type of device is typical of IDT's design approach — integrating added functionality (both memory and logic control) onto a single chip.

Before long, the company was the world's leader in communications memories. Led by two broad families of products — FIFOs and dual-ports — communications memories offer clear testimony to the company's leadership claims, delivering the fastest speed, highest density, widest interface and most extensive set of features. It's why many of the world's leading vendors rely on IDT's parts for their switches, routers, cable modems, cellular base stations, and SONET/ATM multiplexers.

As the communications revolution gained momentum in the mid-to-late 90s, IDT was hard at work delivering a steady stream of innovative, cutting-edge communications products that would enable its customers to add increasing value to the systems they brought to market. To address the needs of the global network, IDT's silicon solutions were targeted at corporate LANs, storage networks, the Internet and broadband technologies, and growing wireless networks.

Today, IDT has a broad portfolio of key silicon innovations:
• communications memories include FIFOs, multi-ports, ZBT and QDR™ SRAMs;
• communications ASSPs include ATM switches, TSI/TDM switches, high-speed PHYs and embedded processors;
• high-speed SRAMs;
• and high-performance logic and clock management products.

IDT's Hillsboro, Oregon, manufacturing facility supports the 0.18-micron leading-edge process technology.

In March 2000, the company took additional steps to better focus its core competencies on creating whole product solutions for the communications market — a market forecasted to grow nearly 30 percent each year for the next five years. It opened a remote design center near Dallas, Texas, and created the Internetworking Products Division (IPD) to accelerate the development of communications-specific products for the exploding voice, data and wireless markets.

According to Dave Côté, vice president of communications ASSPs at IDT, "As the communications market continues to boom, we'll keep refining and evolving our strategy of supplying high-performance, highly featured products to leading-edge communications customers."

To help IDT achieve its goals and enhance its advantages in global markets, IDT has continued to hone its manufacturing strategy. The company's fabrication plants (multibillion-dollar factories referred to as "fabs" in the business) in California and Oregon, along with assembly and test facilities in Malaysia and the Philippines, give IDT the unique ability to combine functional design with manufacturing strategy — using the right process for the right product. Just this year, IDT announced a major manufacturing milestone by moving its manufacturing facility in Hillsboro, Oregon, to the 0.18-micron process.

This ability, along with a fanatical devotion to quality, has put IDT into an elite group of suppliers on the U.S. military's Qualified Manufacturers List and led to the company receiving the Self-Audit Certification Award, STACK International's highest level of quality compliance. IDT has also achieved ISO 14001 certification in fiscal 2000.

The company ended its fiscal 2000 in March, reporting its best revenue numbers to date and growing operating profits dramatically. With revenue of more than $701.7 million, up 36 percent from 1999, and net income of $1.32 per share, the company reported that its gross profit margins had improved for the seventh consecutive quarter — surpassing the 50 percent mark. According to Taylor, "This has been a year of great progress for IDT and its shareholders. We believe the great strides we made are the result of our focus on high-growth Internet and wireless infrastructure segments and on our strong customer-focused business model."

Today, with more than 4,800 employees worldwide serving customers in more than 40 countries, IDT's chips are at the heart of the communications revolution.

Taylor, a 25-year semiconductor industry veteran, has led the company's strong focus on the fast-paced communications marketplace.

Linear Technology Corporation

BACK IN 1981, AT THE DAWN OF THE DIGITAL AGE, MOST ELECTRONICS MANUFACTURERS WERE PREDICTING that digital electronics would soon finish off the analog chip market. In that year, five analog heavyweights from National Semiconductor jumped ship and, armed with $5 million in venture capital, founded Linear Technology Corporation. Nineteen years later, Linear Technology was ranked 18th on average shareholder return over a 10-year period by the *Wall Street Journal*.

In retrospect, the digital revolution not only required analog technology, it ended up requiring analog chips that had not yet been designed when Linear Technology was founded. Linear integrated circuits (ICs) operate in the analog world, where the flow of information and control is continuous (e.g., a wall thermostat or a light dimmer switch). Digital circuits, on the other hand, process information discretely (i.e., in Boolean 1's and 0's). Although digital ICs were to become the showpieces for many Silicon Valley firms, linear ICs were to play equally essential roles in the digital revolution by ensuring that power and information were received by the digital circuit for processing and that the results were returned to the operator. The digital chips inside the early Apple computers, for example, were served by numerous linear circuits of varying sizes and capacities. And, it seems safe to say, there never could have been an Internet without analog ICs because the routers and switches that form the backbone of the Internet are loaded with analog ICs.

Robert H. Swanson Jr., Chairman, CEO & Founder of Linear Technology Corporation

At the end of 1999, two of the company's founders, Robert H. Swanson, Chairman and CEO, and Robert C. Dobkin, Vice President of Engineering and Chief Technical Officer, paused to look back over their company's history. Although neither Swanson nor Dobkin were making any claims about having foreseen the enormous array of new technologies that eventually emerged in the Valley, both were convinced that much of the company's success was due to its having been attuned to three industry trends just coming upon the horizon in 1981: increasing miniaturization, portability and precision, which were to play highly critical roles in the digital revolution.

In many ways the history of Linear Technology mirrors that of the rest of the Valley. When the company achieved $100 million in sales in 1992, 40 percent of its business was with the military. After the fall of communism, the company began looking at the industrial, computing and communications markets as alternatives to selling to the military. Between 1992 and 1994, the market for personal computers (PCs) and portable PCs was beginning to take off. Portable PCs made in those years, including those made by Apple, were loaded with Linear Technology's ICs. By 1996, only 10 percent of Linear Technology's business was then in military products (mostly radiation-hardened devices for satellites), while industrial process control accounted for 35 percent of the company's business, with computers and telecommunications accounting for the other 65 percent (with sales split fairly evenly). By 1999 Linear Technology had achieved over $600 million in annual sales. In the last quarter of that year, communications accounted for 37 percent of the company's sales, up from 11 percent in 1996. Most of this growth was due to the new Internet and wireless markets, as well as to the need for analog ICs in base stations, routers and switchers.

Linear Technology's high-performance amplifiers, voltage regulators, voltage references, interface circuits and data converters were uniquely poised to serve a diversity of markets including telecommunications, cellular telephones, networking products and satellite systems, notebook and desktop computers, computer peripherals, video/multimedia, industrial instrumentation, automotive electronics, factory automation, process control, and military and space systems.

Swanson and Dobkin attribute the company's phenomenal success to the decision early on to specialize in the high-performance analog market, and to avoid unfamiliar markets. In selecting products, Linear Technology's philosophy has always been to look at what the customer needs, and then to base market price on functional value, not on what Linear Technology spends to make the product. Noting that since its founding, Linear Technology has consistently, quarter after quarter, and year after year after year, experienced higher profits and sales, Swanson said, "As a percentage of sales, we're the most profitable semiconductor company, maybe even the most profitable company on the face of the earth. It's all based on not venturing off into unknown areas."

By the end of 1999, the company had earnings of almost 40 percent in after-tax profits. At the same time, the analog market had grown to over $20 billion. And although Linear Technology's $580 million in sales at first glance seems only a small portion of $20 billion, the $6-8 billion high-performance subset of the analog market that Linear Technology competes in is growing at least 25 percent faster than the overall analog market. This fact allows Swanson to predict with some confidence that the company could easily do over a billion dollars in sales by 2003, given market cooperation.

The company's path to success wasn't exactly etched in silicon, however. There was a time back in 1985, for example, when Linear Technology was spending $200,000 per week and only had $2.6 million in the bank. Everyone in the company had taken a 10 percent pay cut, with the option of taking every other Friday off as compensation. Nobody took their extra days off; instead everyone showed up for work on Friday, and Saturday too. But then the company raised more money, at the same time that sales jumped from $4 million to $6 million per quarter. And as it turned out, this boost was all that was needed to get the company back on track.

For its first 10 years, Linear Technology had depended on subcontractors to assemble and test its products. But it was finding that the majority of its delivery and quality problems originated in assembly. So between 1994 and 1995, the company built its own assembly facility in Penang, Malaysia, which several years later was being doubled in size. The company is planning a third wafer fab site (its second fab in Milpitas), as well as expanding the capacity of its fab in Camas, Washington. The Camas and original Milpitas fabs now build all of the company's wafers; the wafers are then shipped to Penang, Malaysia, for sorting and assembly. Linear Technology's Singapore facility is its major electrical test and distribution center. Singapore tests 90 percent of Linear Technology's products, and delivers 70 percent. The Milpitas site delivers the other 30 percent. The company has design centers in Milpitas, California; Boston, Massachusetts; Bedford, New Hampshire; Raleigh, North Carolina; Colorado Springs, Colorado; and Singapore.

Linear Component Supplier by *Electronic Buyer's News* in 1993. Linear Technology won Instat's coveted Kachina Award for the best financially managed semiconductor company four out of five years from 1993 through 1998. The company was the recipient of the Outstanding Corporate Growth Award for Emerging Companies from the Association for Corporate Growth. And for several years, Linear Technology was named one of the Best 200 Small Companies by *Forbes* magazine until it outgrew the small company category. *Business Week* selected Linear Technology as one of its "Hot Growth Companies" in

Linear Technology Corporation Facilities include (from bottom clockwise): World Headquarters — Milpitas, California; Wafer Fab — Camas, Washington; Test & Design Center — Singapore; and Wafer Sort & Assembly — Penang, Malaysia.

For several reasons, locating the company in Silicon Valley was crucial to its success. First, Silicon Valley was where the magic in technological innovation was happening. Second, there was plenty of venture capital available in the Valley in 1981. Third, a unique and supportive entrepreneurial spirit existed in the Valley that did not exist anywhere else. And finally, the Valley was filled with people with the technical skills that Linear Technology needed. Although the company has appendages all over the world, Silicon Valley is still where the heart of the company is.

Throughout its history, the company has won innumerable industry awards. Linear Technology became the first Silicon Valley company to receive Ford Motor's Q-1 Award in 1988. The company was named a Preferred

1999 and ranked it among its Global 1000. And in March 2000, Linear Technology was named to the S&P 500 big company index.

When asked whether they had any advice to offer young entrepreneurs, Swanson and Dobkin advised founders of new startups to pick out something that they can do better than anybody else, and then be prepared to do it really well. Swanson is convinced that a company should not attempt to be "all things to all people." Dobkin added that it is essential that any startup become a real business, one that makes a profit and that satisfies its customers. Given that Linear Technology Corporation has achieved consistent growth in sales and profits since its stock was first publicly traded in 1986, would-be Silicon Valley entrepreneurs might well take note.

AboveNet Communications, Inc.

A SUBSIDIARY OF METROMEDIA FIBER NETWORK

DAVID RAND IS IN THE BUSINESS OF PLOWING GROUND, PROCURING SEED MONEY AND TOUTING THE benefits of fiber, but he is not a wheat grower. As the Senior Vice President and Chief Technology Officer of Metromedia Fiber Network, Rand grows businesses on the Internet. He co-founded AboveNet Communications in 1996 and nurtured it through to its acquisition by Metromedia Fiber Network (MFN) in the fall of 1999. Today, the 7-year-old Metromedia Fiber Network is "planting" 3.6 million miles of fiber-optic cable in 67 cities worldwide. And, unlike anyone else in the industry, MFN will then lease this dark fiber (unlit strands of fiber) directly to its customers. The result, Rand says gleefully, is that "MFN will totally destroy the price of bandwidth in the local market."

But that is not all farmer Rand has in store for growing business at MFN and AboveNet. Besides leasing fiber, which is fast becoming the lifeblood of all Internet commerce, Rand also provides the acreage in which to grow. By leasing data center space to Internet Service Providers (ISPs) and content providers, AboveNet is uniting the forces that make up the Internet. This close proximity of ISPs to content providers simplifies the transfer of information between the two, thus creating an improved Internet for the consumer.

High above San Jose in the ever expanding labyrinth of chain-link cages that take up the top floor of one of the largest buildings downtown, AboveNet is turning what was once storage space into ground zero for Internet connectivity. The AboveNet Internet Service Exchange (ISX) facilities are world-class, leading-edge facilities and they're sprouting up around the globe. Rows of neatly stacked and caged networked servers form the "Net Above the Net," as it is called by company enthusiasts. AboveNet is the realization of what Sherman Tuan and David Rand dreamed of when they first began discussing the limitations of the Internet in the mid-90s.

The co-founders of AboveNet both agreed that the existing infrastructure of the Internet didn't have the capabilities to grow beyond its original purpose as a research network. Their plans to make the Internet a better place by bringing ISPs and content providers closer together proved to be the key to AboveNet's success. Tuan, an engineer from Taiwan, tapped into his entrepreneurial talents while Rand, who developed protocols connecting modems and ISBNs to the Internet and advanced router systems, provided the technical genius to start the company.

The business has proved to be a lucrative one. But like all farms, AboveNet encountered some stumbling blocks, namely the ability to quickly increase bandwidth capacity by adding more circuits. Rand discovered that ordering a new circuit was easy; unfortunately getting the new circuit turned on could take weeks, even months. This type of delay was not acceptable in the Internet business. Rand discovered the value of dark fiber and began ordering fiber from Metromedia Fiber Network. Rand and Metromedia Fiber Network founder Stephen Garofalo soon realized how the two companies could make the Internet a better place — faster and more cost-effectively. A few months later AboveNet was acquired by Metromedia Fiber Network.

Like the railroad, telephone and cable industry pioneers of earlier days, MFN found that wrangling with logistics, regulations and industry incumbents was its biggest challenge. The fiber-optic cable that Metromedia

Dave Rand, Senior Vice President & CTO of Metromedia Fiber Network

Fiber Network deploys consists of a 432- or 864-strand, fiber ribbon that was first developed by Lucent in the early 1990s. At the time, few people had the need for fiber capable of transmitting a terabit per second of information. A typical phone company only installed 12 fibers in a conduit because anything more was unnecessary. When the Internet boom created new demands, the phone companies had a hard time keeping up. Even so, the idea of a private company installing and leasing its own fiber infrastructure to the public raised more than a few eyebrows.

There was no denying that Metromedia Fiber Network was offering a solution to the long delays and prohibitive cost of obtaining fiber, and had the financial backing in place to pull it off. Rand says it costs MFN the same amount of money (as the telephone companies) to lay the cable, but MFN installs 864 strands of fiber instead of 12. Since MFN leases to individual companies, the cost of construction can be amortized over many fibers, so the cost per fiber is significantly reduced. While it is a significant expense to install the fiber, the return on that initial investment is realized quickly.

The vision to extend the metropolitan fiber-optic infrastructure remained intact despite the initial struggles, and after a year and a half the first MFN 432-strand cable strung through a freshly laid conduit system was sold out. Six months later an additional 864-strand cable also sold out. Installing additional fiber once the conduit system is in place is "no big deal," according to Rand. Each conduit can hold three or four of the 864-fiber-optic cables and MFN is installing between four and 22 conduits in 67 major cities worldwide. Even after the planting is completed, the crop of businesses will have plenty of room to grow.

By acquiring their own fiber via long-term, cost-effective leases, customers can gain the Internet scalability and reliability that was previously out of reach. It also allows large companies that already operate acres of data space to lease fiber from MFN and run it from one of MFN's Internet access points to their existing facilities without the encumbrance of third-party telecommunications companies. Now companies can gain access to the fiber and provision more capacity in an afternoon.

The Internet demands immediate response. When a system is not flexible and is overloaded with an unexpected surge of activity, it fails. This creates an influx of unhappy customers. For example, recently a novella was released for sale on the Internet with the expectation that it would sell a few hundred copies. It sold more than 200,000 the first week. The amount of bandwidth needed to download these copies was more than 1,000 times greater than anticipated. The ability to add additional bandwidth during peak operating times is crucial for companies conducting business on the Internet. AboveNet provides this flexibility.

Rand, who has been in the computer business since the 1970s, retains an enthusiasm for the fruits of his labor as though he just began "growing" the Internet yesterday. He claims he has helped to make 40 millionaires in his company alone, but money is clearly not the motivation behind his dedication. "I've made my money already," he says happily. "I'm doing this because I believe in what I am doing. Ten years ago this business didn't exist and now we are enabling a whole new class of industry to succeed. That is exciting to me."

AboveNet's world-class co-location facilities in San Jose, California

AITech International Corporation

A TEACHER PRODUCES IN MINUTES A MULTIMEDIA PRESENTATION FEATURING TEXT, AUDIO AND VIDEO CAPTURES, using a personal computer. The teacher then shows the presentation to students on a big-screen television that is easily visible from the back row of the classroom.

Family members gather in the living room to decide how they will spend the evening. Using their personal computer or set-top box to transmit sound and images to a big-screen television, they can order a movie, play video games, listen to music, surf the Internet, even sing karaoke.

A doctor studies the results of an ultrasound examination on a large television screen. Because the image is a stable, accurate and complete representation of the data produced by the ultrasound scanner, the doctor makes a critical diagnosis with confidence.

Dr. Michael J. Chen brought his background in industrial computer vision and image processing to AITech International, which he founded in 1987.

These real-life scenarios were undeveloped theories in August 1987 when AITech International Corporation was founded by Dr. Michael J. Chen. Chen brought impressive credentials to his fledgling company, having earned a Ph.D. in information science, an MSEE from California Institute of Technology and a BSEE from National Taiwan University. In addition, Chen possessed an extensive background in industrial computer vision and image processing. He has been selected for "International Who's Who of Professionals" and "Who's Who Among Outstanding Americans."

In its first few years, Chen's new Fremont-based company focused on research and development of visual processing technology such as image capture and processing. In 1989 AITech produced an IBM-AT personal computer-compatible peripheral device capable of capturing single-frame images.

With the growth of personal computer use and the development of the Internet, AITech turned its focus toward a new technology with wide-ranging possibilities — the convergence of personal computers and television. Specifically, AITech sought to develop a technology that would allow personal computer-based images to be displayed on a television screen. The challenges in developing such a technology were profound.

Due to differences between the scanning method, signal content, resolution, vertical and horizontal scan rate, dot pitch and contents used by each medium, personal computer images would flicker or appear unstable when broadcast on television. They also were susceptible to poor resolution and color.

AITech rapidly became one of the pioneers in the scan converter industry, solving the conversion riddles with such proprietary technology as digital video encoding, vertical filters with flicker-free algorithms, digital scalers and color space converters. In 1990 the company produced its first scan converter.

By 1995 it had refined that first converter with generations of improvements, twice reducing it to one-tenth its previous size and cost while upgrading its performance. The scan converter is now a single integrated circuit chip that is marketed worldwide.

The impact on the personal computer user was enormous. The new technology, which made it possible to use television as a display device for the personal computer, transformed the computer. Instead of a boring office tool used in a lean-forward posture, it can now be used as an entertainment medium enjoyed in a relaxed, lean-back position in the home.

The new technology also broadened the computer's impact as a professional instrument, making computer-generated productions far more portable and dynamic. Professional presenters can give computer-generated demonstrations on large television screens, making them visible to audiences in vast conference rooms. Teachers can bring computer technology into the classroom and enhance traditional teaching and learning with multimedia contents.

These TV-out products, which convert personal computer signals to television-compatible format, are only half the technological revolution AITech has helped

spearhead. The other half of the revolution is TV-in technology, which converts television signals to personal computer-compatible format.

Through TV-in technology, PC users can capture full-motion video on their computer monitors. They can edit their own home videos and e-mail the final product to friends and relatives. They can watch television shows on their laptop computers, monitor news programs for events that might affect the economy, or just enjoy Monday night football in a dorm room.

One of AITech's most celebrated innovations is the scan converter board that Acuson Corporation placed in its Sequoia 510 Ultrasound System/C256 Echocardiography system. This design, which enabled doctors to project ultrasound scans onto large television monitors, helped Acuson Corporation earn a silver award in the Designs of the Decade: Best in Business 1990-1999 Awards sponsored by the Industrial Designers Society of America and *Business Week* magazine.

Scan conversion technology can have a dramatic effect on leisure-time activities, too. Through information appliances such as set-top boxes, consumers can consolidate various forms of entertainment components such as stereos, VCDs or DVDs, game stations and computer monitors, making their television the hub of their entertainment/information experience. Scan conversion also allows movies to become as portable as music CDs, played on hand-held DVD players and viewed through lightweight goggles.

Such pioneering efforts helped AITech Corporation gain recognition in 1997 and 1998 as "one of the 500 most influential companies in digital media" as selected by *New Media Magazine*. Yet even as it was receiving accolades for its products of the 1990s, AITech was already looking forward to the 21st century and its ongoing objective to be a leading supplier of highly integrated personal computer and television convergence products.

AITech endeavors to refine its existing products, making them smaller, more affordable and more powerful. It plans to add retail locations in the United States and international markets. It seeks to improve its brand-name recognition through increased marketing and advertisement.

With branch offices in Taiwan and an affiliate company in China, AITech is poised to pioneer the next twist in the technological revolution — the convergence of personal computing, television and the Internet into information appliances. AITech plans to be at the center of that revolution, providing the enabling technology that will make bigger visual display possible on new products such as Web phones, mobile phones, digital cameras, palm-sized personal assistants, even Internet-ready household appliances such as refrigerators and microwave ovens.

In this way, AITech will help enhance the contemporary lifestyle and bring broader meaning to the concept of convergence.

AITech's headquarters are in Fremont, California. The company also has branch offices in Taiwan and an affiliate company in China.

Altera Corporation

ALTERA, THE INVENTOR OF ERASABLE PROGRAMMABLE LOGIC DEVICES AND A LEADING SUPPLIER OF PROGRAMMABLE logic solutions, is celebrating the new millennium with a historic benchmark — its first $1 billion sales year. This threshold testifies to the extraordinary vision of Altera's founders, the strength of Altera's new product designs and its innovations in manufacturing and customer service.

In 1983, four co-workers at an electronic design consulting company called Source III knew firsthand the difficulties involved in developing and procuring custom silicon microchips, which were rapidly becoming indispensable for myriad applications. Engineer Bob Hartman, finance specialist Paul Newhagen, technology guru Dr. Jim Sansbury and marketer Michael Magranet understood the problems faced by a computer or communications equipment manufacturer. For delivery of an integrated circuit with a fixed application — now referred to as an application-specific integrated circuit or ASIC — equipment manufacturers faced a major capital investment and a wait of 16 to 26 weeks.

These men teamed up to found Altera in San Jose, with the goal of creating the world's first erasable programmable logic device (EPLD) while providing unparalleled software and customer support. The Altera solution would help customers reduce their time-to-market, giving them a competitive advantage.

By July of 1984 they had introduced the first EPLD, the EP300 device. Altera now offered standard, off-the-shelf delivery of a microchip that could be programmed, tested, erased and reprogrammed by the customer in-house, using only a personal computer. With its combination of CMOS and EPROM cell technologies, the EPLD provided necessary advances in density, flexibility and performance. Altera also provided its customers with the software, expert training and technical support to ensure that the design process flowed smoothly and efficiently.

The programmable chip was a triple breakthrough — in price per volume, flexibility and time-to-market. Within a few years, with annual revenues surpassing $200 million, Altera moved to increase silicon foundry capacity and access to advance process technology through equity investment with silicon foundry partners such as Intel and Cypress. Later partners included Texas Instruments, Sharp and the Taiwan Semiconductor Manufacturing Corporation (TSMC). More recently, Altera partnered with TSMC to build a state-of-the-art wafer manufacturing facility in Camas, Washington, called WaferTech.

Altera Corporation Headquarters in San Jose, California, at dusk

In 1988 Altera continued its innovations by introducing the first advanced Multiple Array MatriX (MAX®) device family, the MAX 5000 device family. The high-performance, high-density MAX device families provide solutions for a broad array of high-performance applications, including simple PLD integration.

Soon after, in 1992, Altera introduced the first architecture of its Flexible Logic Element MatriX (FLEX®) device family, utilizing CMOS SRAM process technology. By combining the time-to-market advantages of PLDs with the high density, high speed and low cost once associated exclusively with gate arrays, Altera's FLEX devices are the most advanced, high-performance and cost-effective gate array replacements available today.

By offering a broad line of programmable logic devices and easy-to-use development tools, Altera presents customers with solutions for the high-speed, high-density and lower-power applications that drive the industry's growth.

In 1999 Altera introduced a revolutionary new embedded architecture called APEX™ (Advanced Programmable Embedded MatriX), which combines the advanced features of the FLEX and MAX products. Also in 1999, Altera launched its next-generation design tool, the Quartus™ software. In addition, Altera offers an extensive selection of ready-made, pre-tested blocks of intellectual property that allow designers to focus more time and energy on improving and customizing their system-level product rather than redesigning common functions. The combination of APEX architecture, Quartus software, and intellectual property enables designers to integrate multiple complex functions on a single piece of silicon for the first time, saving cost and development time. This solution — the system-on-a-programmable-chip solution (SOPC) — is Altera's vision for the future.

While Altera's SOPC solution is a major advance in the PLD marketplace, Altera also realizes that the product distribution system is critical for its customers. Altera has transitioned from the traditional model of a broad electronics distributorship with many partners to a highly focused distribution model with limited partners. Four hundred highly trained Field Application Engineers (FAEs) employed by those distributors represent Altera's personal relationship with each client. Corporate liaison with the distributors is so close that Altera is even involved in their hiring decisions.

Entering the new millennium, Altera's innovations are being integrated into some of the most powerful applications in the world. Currently Altera products serve over 13,000 customers in three primary market areas: communications, electronic data processing (EDP) and industrial applications. Thanks to the worldwide communications boom — including cell phones, the Internet and cable television — the telecommunications sector accounts for approximately 70 percent of all Altera sales. Geographically, 45 percent of sales are to countries outside the United States, with Japan alone accounting for 20 percent of the annual total.

Altera products are the foundation of some of the most complex wired and wireless networks in the world.

Altera Corporation Headquarters soon after opening in 1997

System-on-a-programmable-chip solutions from Altera are the backbone of the world's most sophisticated communications networks. The programming flexibility of Altera's PLDs makes them ideal for communications systems, which must be able to adapt to rapidly changing standards. Companies such as Nortel Networks, Cisco Systems, Fujitsu, Ericsson, Lucent, Motorola, Nokia, Alcatel, Samsung, Siemens, and Sony use Altera's devices in hundreds of cutting-edge communications products.

The independent research firm Dataquest projects that the $26 billion CMOS logic market, of which 10 percent is programmable, will grow at a rate of approximately 20 percent during the first five years of the 21st century. In 1998, Dataquest ranked Altera fifth among suppliers to the worldwide merchant ASIC market. Through constant innovation, intense corporate focus and personalized customer support, Altera continues to consolidate its position in the industry by virtue of its system-on-a-programmable-chip solutions.

Altera's five-year goal is to ascend to No. 1 in the CMOS logic market — and it is well on the way.

ARCOM

STANDING IN THE DELIVERY BAY BEHIND THE ARCOM HEADQUARTERS, LOCATED IN THE HIGH-TECH corridor of Milpitas, Armando Garcia looks every bit the successful Silicon Valley entrepreneur in dress slacks, a smartly pressed pastel shirt, and full-length chef's apron. Once each month, ARCOM's president and CEO cooks for the entire staff, presiding over a six-foot-tall commercial-grade smoker-barbecue. On this Friday, chicken will be served. Last month, it was pork butt. Garcia talks with almost as much pride about his cooking skills as he does the $61 million company that he and partner Jim Vye built from a desk in the corner of Garcia's dining room.

The grandson of migrant farm workers who picked and hauled their way from South Texas to California, Garcia was born and raised in Gilroy, an agricultural community on the Central Coast (also known as "The Garlic Capital of the World"). A high-school dropout, he joined the Navy and trained as a radar technician. This was his introduction to the world of high-technology electronics and provided the foundation for his return to formal education (San Jose State University) and a career in the emerging high-tech industry of Silicon Valley.

Armando Garcia, CEO, in the Service Lab at ARCOM headquarters

In 1976 Garcia became a senior manufacturing technician at Racal-Vadic, a major circuit board manufacturer. He started on the assembly line testing circuit boards but his natural enthusiasm and growing passion for helping others led him from manufacturing to field service support and ultimately to sales.

Five years later, he accepted a job as an outside sales representative for Moxon Electronics. Moxon promoted Garcia to data communications division manager for Northern California and then in 1984, to vice president of the division for the entire West Coast.

ARCOM is ranked as one of the largest providers of information technology solutions in the United States by *VAR Business* magazine.

In the Silicon Valley, hot young sales executives are aggressively recruited. Likewise, Garcia was offered a high-level position with a Moxon competitor. And as is often the case, Moxon took legal action to block his move, an action that ignited Garcia's drive to build his own company. In 1985 ARCOM was established as a manufacturer's representative firm selling hardware for local electronics companies. During this first lean and difficult year, Jim Vye entered the picture.

An experienced engineer with a bachelor's degree in industrial technology and digital design, Vye began his career at Hughes Aircraft in Los Angeles. He started as an assistant project manager in the Radar Systems Division working on F-14, F-15 and F-18 jetfighter projects. Then as part of Hughes' Management Rotation Program, he became a process engineer in the Electro-Optics Division and later a supervisor in the Sub-Contracts Department of this same division. Vye was responsible for over $50 million of outsourced Hughes products, which gave him invaluable business experience in negotiating terms, conditions, pricing, delivery and government contracts.

In 1980 Vye went to work as a program manager on the Cruise Missile Project for Santa Clara's ROLM Corporation. Finally, in 1983 his career path converged with that of his brother-in-law, Armando Garcia, at Moxon. Moving from engineering to sales, Vye made his mark as Moxon's top sales representative. Then as Moxon weathered tough times, Vye's expertise in manufacturing, marketing, contracts and project management was put to work and he eventually became a partner in Garcia's new venture.

By 1987 the two entrepreneurs had decided to broaden their market in the manufacturing industry. The digital networking revolution, which had been moving along slowly and steadily, exploded with the introduction of Ethernet 10-Base-T technology. Fresh opportunities were plentiful and ARCOM became a legitimate Value Added Reseller (VAR) of networking products. The big order, the single deal that would put them onto firm financial ground, came in August of 1987.

Tandem Computers ordered $100,000 of SynOptics Communications equipment. Garcia, looking back in a 1990 interview, mentioned that not making this sale "might have wiped us out." Instead, the partnership with SynOptics and a later one with Telebit (for high-speed modems) set the stage for ARCOM's rapid growth.

Three years after that first big order, ARCOM was featured on the cover of *Hispanic Business* magazine as the fastest-growing Hispanic-owned business in the United States. The company had grown an astonishing 20,000 percent between 1985 and 1989. Sales exceeded $6 million annually and ARCOM now employed 28 people.

Having expanded into new networking product areas, ARCOM entered 1994 on the leading edge of an emerging tidal wave — the Internet. Garcia assigned a new employee, Michele Tayler, to handle a small resellers market, which was then a secondary focus for ARCOM. Many of these resellers were new businesses called Internet Service Providers (ISPs) and Tayler quickly saw the potential. With the support of Garcia and Vye, and with funding from Lucent Technologies, who also understood the potential of the emerging ISP market, Tayler launched the Starcom branch of ARCOM to focus exclusively on the ISP and reseller markets. In January of 2000, in order to avoid customer confusion, ARCOM removed the name Starcom from the market but continues to provide the same level of service to both the ISP market as well as the emerging Competitive Local Exchange Carrier (CLEC) market.

Standing next to a grill the size of a Honda, waiting for the oak to reach the proper temperature, Armando Garcia is happy but not satisfied. ARCOM is ranked as one of the largest providers of information technology solutions in the United States by *VAR Business* magazine. It has won two Lucent Technology Partner Awards for substantial annual sales growth. And for the third year in a row, ARCOM earned a place on the *San Jose Business Journal's* Private Companies List featuring the top 100 fastest-growing privately owned companies in the Silicon Valley.

Sales revenues have increased tenfold to over $60 million annually and the ARCOM work force, well known for its networking prowess and customer service, is now 90 people and growing. The Service Provider Sales Group growth has been just as dazzling, contributing over 50 percent to the company's total revenues.

Vye is now vice president of business development and chief technology officer while Garcia's focus is on sales. While maintaining their strong product orientation, ARCOM's leaders are expanding the company's service business. Increased emphasis has been placed on new technologies like e-commerce, security, quality-of-service, video conferencing, voice-over IP (VOIP) and multiple data services. Vye and Garcia are targeting $500 million in sales revenue in the not-too-distant future.

While considering this future, Garcia smiles a sly smile and admits, "I think we can do $1 billion in business." That "can do" attitude brought this young man out of Gilroy, California, and made him a top contender in the fiercely competitive network equipment and services business.

ARCOM staff in 1988 — Jim Vye is seated second from left and Armando Garcia is seated second from right in the front row.

Asanté Technologies, Inc.

THE STORY OF ASANTÉ TECHNOLOGIES, INC. STARTED THE OLD-FASHIONED, ICONOCLASTIC SILICON VALLEY way — a couple of men, brimming with ideas and technology that could change the world, but unable to raise adequate funds. Wilson Wong and Jeff Lin set up in an unheated warehouse and worked around the clock on borrowed equipment to develop networking technology. Focused on a specific untapped service, or a niche market, Asanté achieved success through its Ethernet, or high-speed networking solutions, with products such as switches, hubs and adapter cards. Using their transatlantic connections, the company was eventually financed by overseas venture capitalists, and Asanté broke even after a mere 18 months and an outlay of only $1 million.

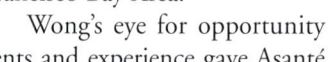

IntraCore 8000

Wong, who holds several patents in computer and design work, designed the world's first network chip, and soon after, IBM's first network adapter card for personal computers. When Wong started his network company for the burgeoning Apple Macintosh market in 1989, he applied his network knowledge to design the first 32-bit Ethernet adapter for Apple Computers. With this humble beginning, Asanté went on to become the leading provider of Ethernet networking solutions for Apple. By December of 1993 Asanté held its initial public offering (NASDAQ:ASNT) and was named the 19th fastest-growing company in the San Francisco Bay Area.

Wilson Wong, Chairman & CEO

Wong's eye for opportunity and his technological talents and experience gave Asanté an edge on competitors for this niche market. But Wong had another advantage — contacts in Asia. He had taught computer science and business administration in Taiwan, and many of his students went on to become presidents of their own companies. Asanté was able to marry the high technology of Silicon Valley with high-efficiency manufacturing in Asia, keeping product costs down. Asanté continues to design its products in San Jose and manufacture in Taiwan and other Asian countries.

However, no company can stand still, according to Wong. Any enterprise in Silicon Valley that wants to be an innovative trendsetter has to operate in a startup mode, or operate in a sustained mindset of intensity to create, produce and deliver unique products or services to market. So in 1999 when the emerging Internet opened up many possibilities for small companies like Asanté, Wong recognized an opportunity to reinvigorate the

company. The growth of the Internet market made time an increasingly valuable commodity as consumers wanted information from the Internet with accelerating speed. To meet that need, Asanté developed high-speed gigabit switches as part of its Internet connection and networking solutions. The company reorganized into two groups to focus on different customers' networking needs. The Advanced Systems group serves enterprise customers and Internet Service Providers (ISPs), and the SOHO (small office/home office) group provides networking solutions for small offices, homes, schools and pre-press markets. Asanté puts it energies into creating the next generation of technology that makes networking easy, and it is a powerful, consistent performer in its market.

IntraCore 9000

Asanté products include a variety of switches (10/100/1000Mbps), hubs, bridges and network adapter cards for local area networks and network management software. Its high-end Gigabit Ethernet multiservice network switches, IntraCore 9000 and IntraCore 8000, based on Asanté's recent IntraCore™ Architecture, serve enterprise network customers and ISPs who have converged telephony, video and data networks. These systems meet the requirements for multiservice networks that support all applications and data types. In addition, the IntraCore design, IntraCore 9000 and IntraCore 8000 switches, gives the customer the flexibility to upgrade the system with new features and meet industry standards without having to invest in new hardware.

To satisfy the needs of home and small-office users, Asanté manufactures a line of FriendlyNET™ products. This line consists of Gigabit Ethernet hubs and switches, network adapter cards, USB hubs, and in the future, wireless and phone wire network products. They can connect PCs, Macintoshes, workstations, network printers and other network resources. They feature a small, compact design suitable for desktop or wall mounting. As the name FriendlyNET implies, these products are designed with ease of use in mind. Plug-and-play features on the product allow a customer to plug it in, turn it on and use it immediately.

Wong takes pride in Asanté's innovative products, which are available internationally through distributors, mail order and computer retail stores. In addition to Asanté's quality products, he believes wholeheartedly that customer care contributes to the company's success, and therefore, Asanté offers extended warranties and free technical support. He also feels gratified about the tangible results of the company's products. For example, Asanté's FriendlyNET systems for small office/home office users assist people in telecommuting and working from home, which in turn eases congestion on the highways.

Although Wong is a man of many degrees, including master's degrees in electrical engineering, business administration and theology, he received perhaps his best education when he worked at Hewlett-Packard in the early 1970s. He witnessed good organization, learned effective management skills and captured a broad understanding of how a successful company operates. And the value of the human aspect in business impacted Wong as he remembers Bill Hewlett grilling hamburgers for his employees at a company picnic.

For Wong, the value of his company is its innovations, and the most valuable asset is his employees. His philosophy is to care for his employees. Company activities are family-oriented, and employees enjoy flexible time and a casual atmosphere. English classes are offered to help new international employees adapt to the United States more easily. Promoting education in the community is also an important cause at Asanté. The company donates equipment and offers substantial discounts on networking equipment for local schools. Wong wants Asanté and his employees to be successful so they will have time to participate in community activities of their choice.

In a joking manner, Wong summarizes his management philosophy in five steps. He says that every day he pats a few shoulders to encourage staff and give positive feedback. He tells a few jokes to keep them happy. He makes a few phone calls to develop and maintain relationships. He surfs the Internet to know what's going on in business. And there's always prayer — especially if the first four steps don't work.

USB 7 Port Hub

Friendly Stack Stackable Switch

10/100 Fast Ethernet Adapter

Atmel Corporation

IN A LAND OF CHEST-BEATING CEOS AND OVERSTATED PROMISES, ATMEL CORPORATION IS A QUIET SUCCESS. Now a billion-dollar-plus company, Atmel reflects the personality and sensibility of its founder, George Perlegos — quiet, determined, results-oriented.

Growing up on a farm in Greece, Perlegos learned a simple philosophy, one that is reflected on the sign outside his office: "Work hard. Make Money." The Perlegos family emigrated from Tripolis, Greece, to Lodi, California, when George was 12 years old. The fertile land of California's Central Valley held promise and George's family became grape growers. Farming taught young Perlegos the value of hard work. Today, the young man has become one of Silicon Valley's most successful entrepreneurs.

In 1972 Perlegos earned his electrical engineering BS degree from San Jose State University. He earned his MSEE at Stanford University and continued engineering studies at Stanford until 1978. Concurrently he held a full-time job as a design engineer at Intel Corp. Perlegos worked on the Intel team that invented the erasable programmable read-only memory (EPROM) chip. A rather persistent inventor, he soon developed a variation of that original technology, leading to the creation of the electrically erasable programmable read-only memory (EEPROM). From this came the development of the flash memory, which, along with the EEPROM, is a mainstay of today's multibillion-dollar, worldwide non-volatile memory business (non-volatile memories hold information after power is turned off).

In 1981 the entrepreneurial bug caught Perlegos and he became a co-founder of Seeq Technology, Inc. After serving as vice president of engineering for four years, it was time to take another turn. In late 1984 Perlegos founded Atmel (an acronym for Advanced Technology for Memory and Logic). Joined by his older brother, Gust, and T.C. Wu, he found himself starting a brand-new chip company just as the semiconductor industry headed into a deep slump. Venture capitalists responded by saying "No thanks." Funding was nearly impossible for a chip startup, especially one founded by executives who were proposing a radical new idea in chip manufacturing.

That new idea was to design, produce and sell chips without a fab — the multibillion-dollar fabrication factory that is built and owned by semiconductor companies. Today, a few other companies are better known for starting off "fabless," but quiet George Perlegos was one of the technology's founding fathers. The Atmel founders and employees can also take comfort in the fact that their company has outperformed all of those noisier, better-known fabless chip companies.

Still, it took some cash to get started and with venture money so tight, Atmel had to be funded by its own founders. Less than $25,000 and a modest line of bank credit got the company rolling, so some lean times followed. One executive remembers having to get to meetings early. "We only had four chairs. So whoever arrived fifth or sixth had to stand."

Those days seem long ago to the newer employees. Atmel went public in 1991 and has since become the world's largest manufacturer of serial and parallel EEPROMs. Through a strategy of smart acquisitions and steady product innovation, Atmel broadened its technology base. In 1993 Atmel acquired Concurrent Logic, which made user-programmable chips. In 1989 Atmel bought Honeywell's Solid State Semiconductor Division in Colorado, which expanded its product range and brought it into the ranks of chip companies who own their

George Perlegos, President and CEO, Atmel Corporation

manufacturing. In 1995 a similar acquisition of ES2 added capabilities in France. By 1996 Atmel's sales and profits were flying high. Revenues were nearly doubling each year. In April of 1996, the company moved into sparkling new facilities in northeast San Jose, where both the buildings and land are owned by Atmel at the core of Silicon Valley.

A storm came in 1997, driven by the Asian financial crisis. Semiconductor companies across the board were hit hard, with memory suppliers taking the worst of it. Prices crashed, sales slowed, profits fell. But George Perlegos and his team would not roll over.

The company was restructured. Production was slowed. A new focus on system-level integration (what's now called "system-on-a-chip") and advanced technologies was formed. Most painful of all, especially to Perlegos, was a 10 percent work force reduction.

Slowly, steadily, in its own quiet way, Atmel returned to profitability. In March of 1998, Atmel purchased European chip maker Temic Semiconductor (previously owned by Daimler Benz). Aside from contributing increased revenues, the Temic acquisition moved Atmel ahead in Silicon Germanium and RF product development. These new chips are used in a wide variety of familiar products like multiband digital cell phones, automotive airbag systems and remote keyless entry systems.

Also during 1998 Atmel bought Data Communications Technology of Research Triangle, North Carolina, to enhance multimedia and wireless communication product development, further expanding Atmel's markets. By mid-year 1999, Atmel was back on track, becoming the profitable $1.3 billion company it was in 1999.

The well-thought-out combination of in-house R&D, acquisitions, and technology partnerships has given Atmel an unusually broad selection of products and technologies. This, in turn, has led to Atmel's tiny chips being placed at the heart of giant-sized market opportunities like cellular phones, smart cards and security systems.

In May of 1999, Microsoft announced that it will use Atmel's family of crypto-controllers in its "Smart Card for Windows" operating system. Market analyst firm Dataquest estimates that all new PCs will come with Smart Card identification as standard equipment by 2002, creating a $6.8 billion market for the chips. Polaroid is also using Atmel's smart card chips to produce digital driver's licenses and document identification for governments around the world.

Atmel now manufactures and ships over 4 million integrated circuits each day — more than a billion per year. Employees number more than 6,000 and operations are spread across North America, Europe and Asia. And the Greek farm boy who started it all and continues to run the show is a Silicon Valley success story, the type of story that is usually shouted from the rooftops.

Not so with Perlegos. Nor Atmel. He's still open, casual and approachable. His attitude formed the corporate personality: if something has got to be done, go do it. Bureaucracy and politics are unwelcome. Instead the rule is: work hard, make money.

Using a silicon germanium (SiGe) process, Atmel manufactures chips that enable the higher switching speeds demanded by today's mobile communications.

Atmel's new worldwide headquarters in the heart of the Silicon Valley, San Jose, California

ATMI

ATMI HAS ALWAYS BEEN 10 YEARS AHEAD OF ITS TIME AS A DEVELOPER OF SEMICONDUCTOR TECHNOLOGY. IN AN industry with a high-speed evolution, ATMI, the leader in front-end materials and systems for the semiconductor industry, is envisioning today the technology for a future decade. The materials and processes it is currently developing will roll into production as new and innovative solutions 10 years from now. Ironically, by then ATMI will have already revolutionized its own novelties. What inspires such prescience? For ATMI, the demand for smaller chips and faster technology are the motivating factors contributing to ATMI's prowess in chemical vapor deposition (CVD), the process of mixing gases to coat thin films on a semiconductor wafer.

However, CVD was a virtually unknown process when ATMI (then called Advanced Technology Materials, Inc.) was formed in 1986. ATMI's initial business became reality as ATMI began its work in a garage and a few trailers behind a gas station. The five entrepreneurs, Gene Banucci, Karl Olander, Ward Stevens, Duncan Brown and Glenn Tom — all chemistry and material science Ph.D.s — had worked together for years. These five men would not only help create a market for innovative CVD materials, but also propel this technology to become the fastest-growing sector of semiconductor manufacturing.

Though the company's products, technologies and services are continually changing, the company's leaders and their vision remain unchanged. It is ATMI's prescient vision that not only produces change for the future, but also provides a thread of stability and continuity throughout the company's history.

VISIONARY IN PRODUCTS AND PACKAGING

ATMI's innovation begins with its customers. It first discovers its customers' unmet needs and then addresses the needs by developing novel products and systems that fulfill these empty niches with productivity-enhancing solutions. The result of ATMI's forward-thinking research and design is a breadth that reflects the company's goal to be a "one-stop shop" for the semiconductor industry.

Though ATMI's newly discovered products are often inconspicuous materials that serve to comprise another more visible product, the company's list of patents is perhaps a more prominent manifestation of its success in innovation. Virtually every ATMI discovery and its resulting product or process initiates one — or a family of — new patents. They have also created an extensive list of company benchmarks. The company's first patent, invented by Tom in 1988, marks the beginning of an impressive list that has been added to every consecutive year since then.

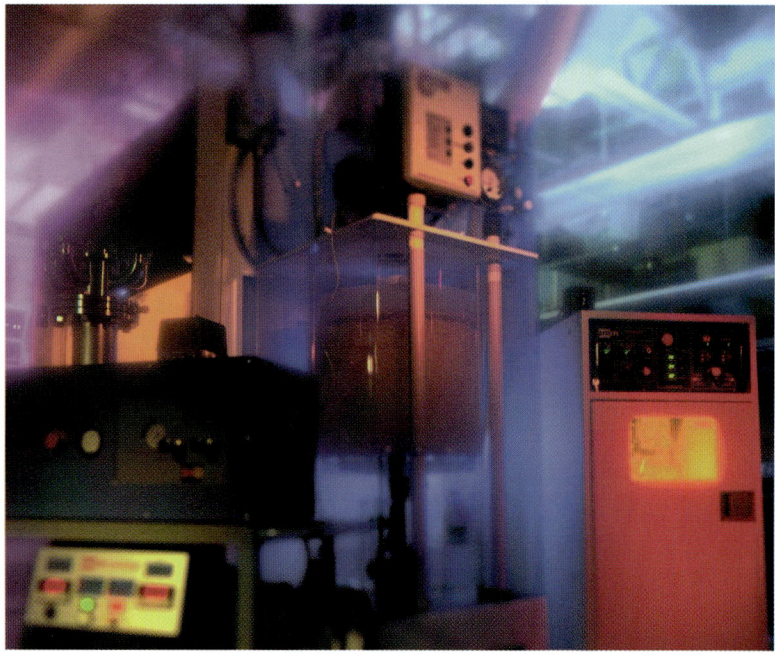

Multiple Treatment Technologies for Semiconductor Processes

Patented innovations are closely linked with the ATMI Ventures group, which focuses exclusively on developing new CVD technologies and materials that address customers' demands. For example, one venture, Emosyn, is a fabless company focusing on non-volatile memory technology for smart card integrated circuit devices.

ATMI also addresses the need for safe packaging and delivery of hazardous materials in the industry. Its Safe Delivery Source (SDS) and Sub-Atmospheric Gas Equipment (SAGE) gas storage and delivery systems have revolutionized semiconductor gas delivery, combining safety with dramatically enhanced productivity. High-purity liquid materials packaging through such innovations as the NOWPak are in widespread use for deep ultraviolet processing and other photoresist materials.

Leader in Environmental Responsibility

While ATMI aggressively pursues innovative materials to enable smaller and faster chips, with equal fervor it completes the materials handling cycle by treating the exhaust materials exiting the chip production process. In the semiconductor industry, chip making requires often hazardous and toxic materials, creating various byproducts. Through its environmental systems business, ATMI addresses the industry's environmental and safety concerns by safely handling process byproducts.

For example, in the early 90s ATMI combined the three critical point-of-use abatement technologies that paved the way for becoming the leader in semiconductor environmental treatment systems. Installing treatment tools directly with semiconductor manufacturing equipment has eliminated many downstream hazardous byproducts. While one method, or scrubbing technique, works effectively on one type of hazardous gas, another technique may be more effective on a different gas. Therefore, to address these varying needs, ATMI created a complete spectrum of services: dry scrubbing, liquid scrubbing, thermal scrubbing and advanced multistage scrubbing. ATMI boasts the industry's only all-inclusive portfolio of abatement techniques.

While ATMI's products, services and customer training programs lead the industry toward higher standards of environmental safety and awareness, its own facilities and operations provide an example of environmental responsibility and leadership. ATMI's technology uses the least amount of water in the industry, and its products are geared toward minimizing water usage. Some use none at all. At the core of ATMI's environmental division is the safety and health of customers and employees and a sustainable future for the environment.

Innovator of Partnerships

While at the forefront of the industry, ATMI is not alone in its work. It seeks out partners who will enhance its new product offerings and help it successfully enter new market segments. As an innovator of such partnerships in the industry, where front-end manufacturers and suppliers collaborate, ATMI has not only built its own market strength but also encouraged research and development that benefits the entire industry.

Founders and original staff in Danbury, Connecticut

One example from the early 90s was ATMI's leadership of the DRAM Consortium in alliance with IBM, Texas Instruments and Micron Technology. The group, funded by both the U.S. government and the alliance members, created new materials for 1-gigabit memory chips and also led to additional partnerships and developments. ATMI expanded its expertise and alliances in these high-k and ferroelectric areas with such companies as Lucent Technologies and Siemens to help improve the performance and reduce the complexity of cellular technology. A partnership with Praxair resulted in new methods for recycling PFCs.

Whether developing its own new processes and patents or collaborating with others in strategic partnerships, ATMI is helping to set the pace for new semiconductor materials technology and semiconductor systems while actually shrinking the development time of future technologies. From initial thin film development to the packaging and transfer of products to the development of safety and environmental standards, ATMI offers a complete portfolio of products and services while continuing to transform the semiconductor industry.

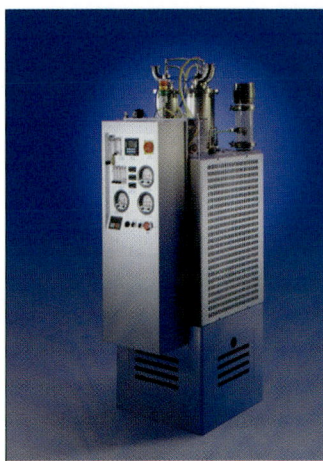

(Far left)
PFC Treatment System using Advanced Plasma Technology

(Left)
Treatment System for SiN Semiconductor Processes

BEA Systems, Inc.

Bill Coleman, chairman and chief executive officer of BEA Systems, Inc.
Photo by Monique Schoenfeld

VISIONARY BILL COLEMAN AND TWO COLLEAGUES, ED SCOTT AND ALFRED CHUANG, DID NOT NEED A CRYSTAL ball to see the future of the World Wide Web. All three men were working for Sun Microsystems when they witnessed the emerging client/server technology revolution in 1993 and 1994. They soon realized that this network wasn't going to support a vital component of the industry — e-business and e-commerce software and its applications.

> **BEA's wide range of education, customer support and professional services helps small companies and large corporations get e-commerce initiatives off the ground and up on the Web.**

Another revolution needed to take place, one of network-based software providing online, Web-based business transactions. Today, the Web has evolved from a convenient informational tool to a valuable resource for all kinds of transactions, especially in business and commerce.

To Coleman, Scott and Chuang it was clear that businesses would increasingly need to provide an easy, safe and prompt way for e-generation businesses and customers to conduct business online, and even more important, that those organizations would need to build new systems that worked with existing systems, integrating and upgrading applications to take advantage of the Web. Seizing the opportunity to be the first to take advantage of a new approach to e-business operations, the three men formed BEA Systems, Inc., the name being derived from the first letter of each man's first name. The founders wrote a business plan in 1994, obtained a small seed grant from a leading venture capitalist, and began focusing on acquiring key pieces of the software to build their transactions platform. When they learned that Novell's Tuxedo® software could do the job, they purchased it from Novell; next they transformed Sun Java architecture into a vital component of their transactions platform.

BECOMING THE E-COMMERCE TRANSACTIONS COMPANY™

The e-commerce market took off in 1995, the same year BEA was founded. By the summer of 1996, BEA had completed seven acquisitions and was growing dramatically in the United States, Europe and Asia. The company boasted $65 million in sales and 500 employees when it went public in April 1997. In 1999 BEA transformed itself from a middleware company into the E-Commerce Transactions Company™. BEA claims a 70 percent market share as a provider of e-commerce infrastructures largely because of the success of BEA Tuxedo and BEA WebLogic® application servers, the flagship products of the BEA E-Commerce Transaction Platform™. These comprehensive e-commerce servers allow businesses to rapidly build, deploy and manage e-commerce applications.

The BEA E-Commerce Transaction Platform also includes reusable components (i.e., building logic that can be snapped together into applications within weeks), integration technologies and services. Introduced in February 2000, BEA WebLogic Commerce Server™ combines enterprise Java components with the BEA WebLogic application server to enable companies to rapidly build and deploy adaptable and personalized e-commerce applications. Additionally, BEA helps integrate companies inside and out. BEA eLink™ integrates new Web applications with existing enterprise systems, and BEA's new E-Market Integration (EMI) technology — BEA WebLogic Collaborate — enables businesses to set up collaborative e-markets with their trading partners quickly. Finally, BEA's wide range of education, customer support and professional services helps small companies and large corporations get e-commerce initiatives off the ground and up on the Web.

Since BEA's original technology was developed by and purchased from Bell Laboratories, the systems have the same high reliability and security of a telephone system. Although computers and networks can fail, BEA's Web applications software is adaptable and can move over to other networks and add new computers when other computers fail. Businesses have come to rely on the BEA E-Commerce Transaction Platform because it performs flawlessly around the clock. It has been tested in the most challenging circumstances and is proven to work, without fail, no matter how many thousands of transactions are processed. BEA believes the successful completion of every transaction will become the most critical success factor in its industry. This reliability of the BEA E-Commerce Transaction Platform has become the company's hallmark and is credited for much of the company's success. The software also implements multiple inherent security technologies and can be configured to provide very secure customer access.

Meeting the Demands of the E-Generation

Today, BEA has more than 4,000 customers worldwide, including leading e-commerce companies such as Amazon.com, FedEx and Wal-Mart. Amazon.com chose the BEA E-Commerce Transaction Platform to help its customers easily shop for an increasing variety of products over the Web. The platform enables Amazon.com to develop, deploy and manage its applications more effectively while providing the capability to support continuous changes in its business. United Airlines relies on the BEA E-Commerce Transaction Platform to provide a constant feed of information — aircraft schedules and ticket prices — to 30,000 ticketing stations worldwide, keeping ticket counters updated and planes on schedule.

FedEx uses BEA software for its package tracking and logistics systems (and some 30 other applications), handling an average of 150 million transactions daily

> **BEA claims a 70 percent market share as a provider of e-commerce infrastructures largely because of the success of BEA Tuxedo and BEA WebLogic® application servers, the flagship products of the BEA E-Commerce Transaction Platform™. These comprehensive e-commerce servers allow businesses to rapidly build, deploy and manage e-commerce applications.**

that track 3 million packages delivered to 211 countries every weekday. TRIP.com's intelliTRIP provides travelers with a fast and easy-to-use tool to search for the best airfare and to purchase tickets directly from the airline's Web site. intelliTRIP relies on the BEA E-Commerce Transaction Platform for this service. Other prominent applications of the BEA E-Commerce Transaction Platform include online banking and auctioning and allowing telephone companies to offer DIRECTV's satellite service when they hook up their customers' phones.

BEA and the E-Commerce Future

Due to recent rapid growth, the San Jose-based company realized $469 million in revenue in 1999 and presently has 2,300 employees. Looking ahead, the BEA vision is to continue transforming the old economy to the new economy by providing one of the key infrastructures that support these changes. Serving as the backbone of an evolving technological marketplace, BEA will continue to provide its clients with the tools necessary for increased productivity and unprecedented success.

Cadence Design Systems, Inc.

Ray Bingham, president and chief executive officer of Cadence Design Systems, Inc.

A young girl prepares for a soccer competition during the 1999 San Jose Inner-City Games, a charitable organization sponsored by Cadence.

HIGH TECHNOLOGY, MICROCHIPS AND SEMICONDUCTORS ALL COME TO MIND WHEN THINKING OF SILICON VALLEY. But artists? Hardly! While all the invention, calculation and atomic bombardment is going on, someone has to decide what the next generation of technology masterpieces will look like and how they will function. It is the design engineers, or high-tech artists, who pull the form and function together into cohesive chips, printed circuit boards, and systems. Cadence Design Systems, Inc. provides these electronic artists with the tools to realize the technology innovations for which Silicon Valley is renowned.

Cadence is the world's leading provider of electronic design automation (EDA) software and services. Its solutions are used by companies across the globe to design and develop state-of-the-art electronics. Many of the products used in everyday life, or the electronic devices used to power these systems, were designed using Cadence tools.

EDA tools have been around for some 30 years. However, it was during the electronics revolution of the 1980s that the commercial EDA industry and Cadence emerged with new innovations. In 1982 Dr. Alberto Sangiovanni-Vincentelli and Dr. Richard Newton, both professors at UC Berkeley, decided to start a company to support some of the software tools developed there and to build the foundation of a new approach to integrated circuits. At the time, both men were developing new databases, graphical editors, algorithms and tools for layout automation and synthesis, and were working toward the idea of an open system to support EDA tools. They met Jim Solomon, who was using the Berkeley software at National Semiconductor. He accepted their offer to head up the new company. Four corporate partners were found to provide the bulk of the start-up funding. First called SDA Systems, for Solomon Design Automation, the company merged in 1988 with ECAD, Inc., another EDA pioneer founded by Paul Huang, who was in contact with the founding team of SDA since its early stage. The merger resulted in Cadence Design Systems.

Their technical breakthroughs formed the foundation of a visionary company that has helped reshape the electronics industry by breaking new ground with more than just the innovative design automation software that its engineers produced to simplify and automate complex tasks. Led by an energetic and ambitious CEO that the founders appointed — Joe Costello — Cadence pioneered new approaches in what was generally regarded as a sleepy industry. A rapid succession of mergers and acquisitions catapulted Cadence up the industry ranks. By taking calculated risks with new business models, such as a "software-only" model (as opposed to competitors' turnkey systems comprised of workstations preloaded with the software tools), and innovative pricing and licensing agreements, Cadence challenged the norm.

By the end of its first decade, Cadence was the world's top EDA powerhouse. But that wasn't enough. In the early 1990s, Cadence set a new course for the industry by launching the world's first chip design services business, culminating in its 1995 landmark agreement wherein Cadence took over Unisys' chip design operation, a first in the electronics industry. Cadence has partnered with many companies and countries ensuring its continued success. Whether a nontechnical company needs a chip designed, or a seasoned electronics manufacturer needs additional engineering resources or specialized expertise, Cadence fills the need.

Under the leadership of Ray Bingham, president and CEO, Cadence has successfully entered the design and methodology services arena, further enhancing its customers' productivity and profitability. Cadence now has the world's largest independent design services organization and the broadest portfolio of technology-based software products in the marketplace. With more than 5,000 employees — in sales offices, design centers and research facilities in Japan, Asia Pacific, Europe and the United States — Cadence possesses expertise in all aspects of electronic product development and design. Approximately 1,500 of those employees are designers who provide advanced electronic design software tools and time-proven methodologies. The design services organization creates world-class product and component designs for a broad range of Information Age companies. Cadence is the only EDA company that devotes significant effort to basic research in its Cadence Berkeley and European

(Far left)
The entrance to the Cadence campus before its redesign

(Left and below)
The new look at Cadence's 11-building campus

Laboratories. There, scientists have been developing software solutions for a future generation of electronic products. Fundamentally, software innovation built the company and remains a cornerstone of the Cadence legacy and lasting impact on the worldwide electronics industry.

As the millennium neared, Cadence became the first EDA company to break through the billion-dollar revenue mark in a single year with 1998 revenue totaling more than $1.3 billion. It has reinvented itself twice to position the company for further success in an ever changing industry.

At the heart of the company is a culture of innovation and leadership — qualities that serve its customers, employees and the Silicon Valley community well. Headquartered in San Jose, Cadence occupies a sprawling 1.1 million-square-foot, 11-building campus. The newest addition, completed in March 1999, is the Cadence Executive Briefing Center, a state-of-the-art presentation and conference facility. The Cadence workplace has come a long way since the company's early days. Back then, company founders say, they could often hear the screams of people on park rides at Paramount's Great America theme park nearby.

Today, Cadence employees enjoy an on-site health and fitness facility, complete with the latest exercise equipment and professional trainers and instructors. There are on-site dental, dry cleaning and car-care services, as well as company-sponsored holiday gatherings for employees' children. Employees enjoy concierge service that provides personal assistance in planning events, making travel arrangements, performing minor automobile maintenance and securing baby-sitting services.

Having enjoyed years of rapid and profitable growth in the Silicon Valley, Cadence also takes very seriously its responsibility to the local community. In October 1999, the nonprofit organization Project HIRED honored Cadence as a corporate partner when the company donated more than $25,000. The same year, Arnold Schwarzenegger, chairman of the national Inner-City Games Foundation, presented Cadence with a plaque commemorating the company's ongoing assistance with the San Jose Inner-City Games. Cadence's involvement at all levels of this event enabled 400 at-risk children to receive training and mentoring at a computer camp and to enjoy participation in a variety of sporting events.

Cadence has taken the lead role in the Silicon Valley's annual Stars and Strikes bowling tournament. This fund-raising event — created, coordinated, and underwritten by Cadence in partnership with nonprofit foundations — has raised more than $2.2 million since its inception in 1990. Benefactors have included the Bill Wilson Center, San Jose Unified School District's Homeless Assistance Program, Atypical Infant Motivation, Inc., Santa Clara Valley Youth Foundation, Santa Clara County Library Internet Project and others.

Cadence's mission is to help companies exploit and leverage the power of electronics. In a world increasingly dependent on high technology, the company plans to expand its horizons with the addition of electronics business consulting. Company officials say that will enable Cadence to forge relationships with local governments and nations, as well as consumer, industrial and light industrial companies looking to integrate the electronics into their next-generation products. While what Cadence does is technically challenging, it offers a simple yet powerful value proposition once within reach of only the electronics elite. The ability to conceive and deliver competitive electronics-enabled products to market faster, with the right features and at the right price, is the key to success in today's global marketplace. For Cadence, making dreams a reality for companies of all kinds, of all sizes, and in all regions of the world is the challenge and opportunity of the 21st century.

Cisco Systems, Inc.

Cisco President and CEO,
John Chambers

John P. Morgridge,
Cisco Chairman of the Board

IF JOHN CHAMBERS HAS HIS WAY, CISCO SYSTEMS, INC. WILL BE THE WORLD'S FIRST TRILLION DOLLAR COMPANY. And if that's not enough, he'd also like Cisco to go down in history as the business that led the Internet Revolution. But, lofty goals notwithstanding, Chambers, President and CEO of Cisco Systems, Inc., has already proven his company's worth. In the first quarter of 2000, Cisco found itself vying for the position as the world's most valuable company, challenging business icons like Microsoft and General Electric.

The founders of Cisco, though enterprising, did not set their sights so high. In 1984 Len Bosack and Sandra Lerner, a married couple working in separate departments at Stanford University, wanted a system for communicating with one another through their computer networks on campus. Putting feet to their passion, Bosack and Lerner built the first multiprotocol router, a computer that allows computer networks to talk to one another.

Today, in an increasingly Internet-driven economy, Cisco supplies 80 percent of the routers over which virtually all Internet traffic travels, making Cisco the worldwide leader in Internet networking. Cisco solutions not only connect people to the Internet, but also create the system that connects everything with high-speed "pipes" — computers and telephones, mobile devices, home appliances and even automobiles. Through such efficient and speedy systems, Cisco offers its customers competitive advantage, cost savings and the opportunity to build closer relationships with their clients, prospects, business partners, suppliers and employees.

Cisco's successful evolution into a leading company is partly due to wise acquisitions. Between 1993 and 2000, Cisco completed 54 acquisitions, 24 of which occurred during 1999-2000. When acquiring other companies, Cisco looks not for businesses with competing products but for those with similar corporate cultures and values that would add to and complement Cisco's portfolio. Each acquisition brings new vision into the company that ultimately enables Cisco to create new products.

KEY TO SUCCESS: VALUING BOTH CUSTOMERS AND EMPLOYEES

Most important to Cisco's success are its customers and its employees. Cisco's number-one priority and passion continues to be the customer. Cisco is committed to helping its customers become agile by implementing Internet business solutions that will position them for success in the fast-paced business environment. Cisco's leading-edge customers deploy networks that deliver a variety of media over the Internet, and they repeatedly rely on Cisco's expertise in providing combined data, voice and video networks.

In the mid-90s, Cisco predicted that the Internet would change the way people work and communicate, personally and professionally. Today few people will argue with Cisco's prediction. But even more important, that prediction also guides the company's philosophy, its corporate culture and its leaders.

Cisco's executive staff is in constant communications with company employees. John Chambers sponsors monthly breakfasts that honor employees celebrating their birthdays. He plays host to employee question-and-answer forums. Quarterly managerial meetings provide opportunities for unifying the company's goals and strategies, which is no small task considering there are 12,500 Cisco employees in the United States and 29,000 worldwide. Cisco Chairman John Morgridge passionately promotes Cisco culture, which is built on the quality of the Cisco team, the drive to stretch goals, and a relentless focus on customer satisfaction by staying in close touch with its clients.

Chambers and Morgridge have a revolutionary approach to business. They listen to both employees and customers. This creates a state of perpetual revolution, but employees find that they want to push "outside the box" to find innovative, creative solutions. They are given the freedom to modify or redefine traditional processes and procedures, for any Cisco employee who has a new vision or concept will find support to pursue it. This open corporate culture resembles that of a fresh startup company where every employee helps define the company's future and directly benefits from its successes.

Cisco's approach to customer service is likewise unique. By listening to all customer requests and not favoring one technology over another, the company can provide its customers with a range of options from which to choose.

The handsome campus of Cisco Systems, Inc., San Jose's largest employer

CHANGING THE WORKPLACE

The range of solutions created by Cisco is broad and deep, leveraging the Internet to dramatically transform the workplace. Cisco, for example, has applied the Internet's power to its "virtual close," an application that allows a company to streamline its financial operation by closing its books in one day. Another application, "e-manufacturing," allows Cisco to seamlessly manage 37 key manufacturing suppliers, partners, and distribution services over the Internet without employee involvement. Cisco's own success with virtual close and e-manufacturing are models that its customers may adopt to reduce overhead, connect all their systems, manage inventory and operate more efficiently, cost effectively and competitively.

CHANGING LEARNING AND LIVING

As an industry leader, Cisco recognizes the importance of its continued participation in the global community to help improve people's lives. In a unique partnership with the United Nations Development Program, Cisco has undertaken an effort to end extreme poverty in the world's poorest nations through a program called NetAid.

Cisco also teams up with educational institutions worldwide to ensure that today's students master the necessary technological skills for success in the Internet economy. The Cisco Networking Academy Program, an e-learning solution with over 3,600 academies in 61 countries, prepares students to obtain a higher degree and to succeed in the workplace upon graduation. The program, which increases the pool of qualified information technology professionals, is designed to be effective in all geographic areas. At Cisco, the philosophy of changing learning, working and living has only just begun.

THE PATH FORWARD

The Internet Revolution seems destined to have as big an impact on nations as that created by the Industrial Revolution. The immediate future is saturated by optical and wireless technologies — the entire networked economy is all about speed and continuous change. Cisco is extremely proud to play a key role in leading the Internet economy. As the company moves forward, it will continue to gauge its competition with caution, make acquisitions that will complement the company's goals, strive to employ the brightest workers and ultimately work to make a difference for all humanity. Together with its shareholders, customers, employees, partners and suppliers, Cisco is impacting the face of the global economy as the entire world moves into the Internet century.

Flextronics International

WHEN JOE MCKENZIE LAUNCHED FLEXTRONICS IN 1969, HE PROBABLY DIDN'T HAVE A $6 BILLION GLOBAL company in mind. In fact, the thought of a "board stuffing" business outgrowing annual revenue of more than a few million dollars was unheard of 30 years ago. But today, Flextronics is one of the world's leading electronic manufacturing services providers building complete products that range from complex printed circuit board assemblies for computer workstations to personal digital assistants.

A manufacturing engineer by trade, McKenzie started Flextronics after the company he worked for went out of business. He and his first employee, his wife, provided an overflow manufacturing service to Silicon Valley companies. At the heart of most electronic products is a printed circuit board, an insulated shelf upon which semiconductors of various shapes, sizes and functions are mounted. The connecting circuitry is manufactured into the board so that once the parts are mounted, the assembly becomes a working device. Companies that needed more boards than they could produce in-house sent their overflow work to Flextronics, where the McKenzies hand-soldered all the parts onto the boards and then returned the finished goods. Since the parts and boards were supplied by the customer, companies like Flextronics were called, somewhat disrespectfully, "board stuffers."

Disrespect aside, it was a good business, if not a rocket-growth business. So in 1980, Flextronics was sold to Bob Todd, Joe Sullivan and Jack Watts. Todd became CEO and put the company on the path that led to today's multibillion-dollar enterprise.

Todd's actions changed the face of both the company and the industry. No longer would Flextronics be a board stuffer, but instead a contract manufacturing firm — a name that to this day identifies the industry. The company would also pioneer automated manufacturing techniques to reduce the labor intensity (and associated costs) of assembly. Flextronics instituted board-level testing to assure that quality and yield targets were met. And in 1981, the company was the first American manufacturer to go offshore, setting up the Flextronics Singapore facility.

By the mid-80s, Flextronics started to deliver turnkey solutions. Customers would create a product specification and send it to Flextronics. Everything from the manufacturing process to the buying of parts was placed in Flextronics' hands. The customer would validate the plan, but no longer did it dictate the process.

To serve this growing trend, Flextronics offered computer-aided design (CAD) capabilities. Customers could come with a product concept and Flextronics would design and blueprint the entire printed circuit board. As an added benefit, component testing was provided so the individual parts could be assessed and quality assured. Flextronics prospered and in 1987, went public — three weeks before the market crashed.

Flextronics International's San Jose, California, campus is located in the heart of Silicon Valley.

Still, by the end of the decade, Flextronics grew and continued to prosper, eventually moving into "full box assembly," the contract manufacturer's term for building a complete, working, shippable product. Sun Microsystems, famous today for its Internet servers, used Flextronics-built disk and tape subsystems in its workstations. The Hayes modem that enabled people to go online and became a standard in the industry (the term "Hayes-compatible" was used for years) was also a Flextronics-built product.

During these years, Flextronics built a global manufacturing base with complete factories throughout Asia. These facilities were highly capital-intensive and, though profitable, relied on a high-volume U.S. market; a market that crashed during the first years of the new decade. The early 1990s took their toll on countless Silicon Valley companies and Flextronics was not spared any of the pain. Profits shrunk and losses mounted. Survival became the target.

Because the Asian operations were still profitable and the U.S. operations were unprofitable, Flextronics management wanted to scale back or close the U.S. plants. But closing a manufacturing facility is an incredibly expensive undertaking, one that would surely bankrupt the company. So instead the Asian plants were spun off as a separate company, outside funding was acquired to "take the company private," and the U.S. plants that were pulling the company down were closed. The complexity and success of the deal could be taught in business schools. Suffice it to say that 13 law firms were involved. And Flextronics survived.

The "new" privately held Flextronics was based in Singapore — still the worldwide headquarters. Michael Marks became Chairman in July 1993 and CEO in January 1994. Also in 1994, the Singapore-based company had a successful initial public offering (IPO), becoming a publicly traded company for the second time (NASDAQ: FLEX).

Marks drove a strategy that rebuilt the company's U.S. presence. At the same time, he led a worldwide management team that turned the company around. From a net loss of $6.5 million in fiscal 1992, Flextronics showed a net profit of $6.2 million just three years later. During that same period, revenues nearly tripled, reaching $237 million in 1995. The company was positioned to become a billion-dollar manufacturing firm — a goal that was reached in 1998.

From 1993 to 1998, Flextronics acquired over 12 operations, built a global infrastructure for high-volume manufacturing, expanded purchasing and engineering capabilities, grew from 3,000 employees to over 13,000 and upped the revenue target to $5 billion.

Original equipment manufacturers (OEMs) that outsource their manufacturing rely on the quality and expertise of companies like Flextronics.

One of the driving strategies that has enabled Flextronics to focus on such a high-revenue target is the company's unique industrial park model. Located In Asia, Europe and the Americas, these parks co-locate suppliers on the same campus where manufacturing takes place, resulting in greater operational flexibility and responsiveness to customer needs. Another unique service is provided by Flextronics' Product Introduction Centers. Throughout the world, these facilities design, prototype, test and launch the production of new products, shortening critical time-to-market. 3COM's Palm Pilot was a Flextronics-designed and built product, as was the Microsoft mouse.

In February 2000, Flextronics exceeded $5 billion in revenue and has been recognized as one the world's largest and most profitable providers of electronic manufacturing services. Today, the company offers advanced engineering, design, manufacturing and distribution services to leading original equipment manufacturers (OEMs), including Cisco, COMPAQ, Hewlett-Packard, IBM, Lifescan, Lucent, Nokia, Philips, Sony and WebTV. The little board stuffer is now a worldwide powerhouse. And the revenue targets keep moving higher. But having survived boom and bust, and having demonstrated a 70 percent annual compound growth rate from 1995-1999, Flextronics' targets look less like dreams and more like bulls'-eyes.

GlobalCenter

LEO J. HINDERY, JR., CEO OF GLOBALCENTER, EXPLAINS HIS COMPANY VISION AS "DELIVERING A TOTAL Internet Services Solution to our customers, including the global infrastructure and the performance necessary for them to succeed in the new economy. This includes an international Internet protocol (IP), network-based, information technology (IT) sensitive, strategically supported, complex Internet service." But when asked what all of that means he says, "We are helping to change the world."

GlobalCenter is an Internet services business that provides total Internet solutions with complex Web hosting through its proprietary infrastructure, utility applications and professional services to manage and maintain the magnitude of volume produced by some of the most highly trafficked online enterprises.

Started in the mid-90s as an offshoot of Global Village, a small company that made fax machines and modems, GlobalCenter provided small office/home office business solutions through modem sharing, T1, ISDN services and dial-up on demand e-mail. By acquiring innovative companies such as ISI, which at the time hosted Netscape and Yahoo!, GlobalCenter quickly became a cornerstone of Internet hosting. In 1998 the company was purchased by Frontier Communications, giving GlobalCenter access to its own network. In September 1999, Global Crossing acquired GlobalCenter through its purchase of Frontier, providing the company direct access to the world's first global fiber-optic network.

In the past four years Global Crossing, GlobalCenter's parent company, has grown from 150 employees to more than 15,000 with annualized revenues of $4 billion. The company has expanded its network presence worldwide to include regions from East Asia to Europe, as GlobalCenter is building in excess of a million square feet of data centers to keep up with the demand of some of the largest and most densely trafficked sites on the Web. According to Hindery, some GlobalCenter customer sites get as many as 625 million page views per day.

During the last months of 1999, after nearly a decade as a leader in the cable industry and most recently as President of Tele-Communications, Inc. (TCI) where he created $50 billion in value for its shareholders, Hindery decided he was ready to take on the challenges of the Internet. At GlobalCenter he found what he was looking for: a fast-growing company with unlimited potential operating at the center of a new era in Internet productivity. He is excited by the technological potential he sees in both his clients and staff. This enthusiasm is reflected throughout the high-energy open-office environment of GlobalCenter's headquarters in Sunnyvale, California. Whether it is the lemon-colored walls, glass

Network Control Center at GlobalCenter
Photo by Glenn Steiner

A GlobalCenter data center
Photo by Glenn Steiner

conference rooms or the complimentary beverages and snacks available in the employee lounge, one gleans that GlobalCenter is a pleasant environment in which to work.

The challenge Hindery faces is melding the responsibility of the business with creativity. He says the Silicon Valley is a "left brain/right brain" kind of environment. Understanding that concept is the first step. Managing it is the second. Above all else Hindery is a businessman. He realizes that no matter how exciting or earthshaking new technology is, at the end of the day it must be profitable. Building value for shareholders remains a top priority. More than anything, he likes helping his customers achieve their goals.

By enabling online companies to handle an increasingly large and variable amount of Internet traffic, Hindery sees GlobalCenter as an integral part of the Internet's continuing evolution. As part of Global Crossing, GlobalCenter's access to a network of 101,000 route miles of fiber-optics serving five continents, 27 countries and over 200 major cities gives e-commerce sites the flexibility to expand worldwide.

When *San Jose Mercury News* reported that online retail sales totaled $5.3 billion in the fourth quarter of 1999, Hindery responded by proclaiming that e-commerce is only the "tip of the iceberg" as far as the Internet is concerned.

Hindery explains that the commerce sites attract a lot of attention because they are so visible, but they are a relatively simple use of the Internet. Buying an article of clothing online is a simple task, technologically speaking, compared to the highly interactive Web sites that also take advantage of the larger bandwidth available through companies like GlobalCenter. Today, the Internet is capable of bringing *Gone With The Wind* to computer screens and providing chat rooms where friends and family can converse together. Students have access to long-distance learning Web sites where they can conduct science experiments or examine rare biological species using interactive methods not possible without a "virtual" setting. These highly sophisticated interactive sites are what Hindery is referring to when he talks about changing the world.

In fact, GlobalCenter soon will be hosting the Library of Commerce Web site that will, for the first time in history, enable anyone to take a look at documents such as the Lincoln's Gettysburg Address. Now, visitors to the Web site can examine the envelope where Lincoln first scribbled the words to his famous speech and see first hand what became one of the most seminal pieces of political thought in the history of this country. In an era where facilities-based education often falls short of meeting the needs of children, especially in lower-income earning areas of the country, Hindery believes that hosting this type of Web site can have a profound impact on society.

"I keep telling my employees it's not about zeros and ones going down the pipeline. It's about certain zeros and ones going down the pipeline. I think every day they are more and more aware that what they are doing is changing society."

Rows and rows of racks at the heart of one of GlobalCenter's data centers
Photo by Glenn Steiner

Employees gather in GlobalCenter's colorful lounge.
Photo by Glenn Steiner

Kinetics

High-purity process delivery system

David J. Shimmon, President and CEO, Kinetics Group, Inc.

KINETICS' PIONEERING HISTORY AS A GLOBAL PROCESS-PIPING CONTRACTOR MIRRORS THAT OF ITS Silicon Valley setting. Just as the region's economy originally focused on farm produce, Kinetics began operations as an agricultural-support business. As the local economy gave birth to a revolutionary new industry in hardware and software products, Kinetics evolved in parallel to become a successful enterprise of high-tech products and services, marketing an integrated range of solutions to the semiconductor, biotechnology and pharmaceutical industries from 40 locations worldwide.

FROM FOOD PROCESSING TO PROCESS SYSTEM SOLUTIONS

When Kinetics was founded as a food-processing business in 1972 by William A. Bianco, the company's specialty was a mobile, mechanical system for storing and shipping freshly picked produce — mostly strawberries and tomatoes. Although it took only two years for Kinetics to build and maintain a network of its patented systems from Texas to California, the produce business itself occupied the company only eight months a year.

By the late 70s, the orchards of what would become Silicon Valley and elsewhere were also becoming home to a growing number of start-up technology endeavors. To stay busy during the off season, Kinetics found work on minor projects for up-and-coming companies — such as Fairchild, Intel and National Semiconductor — to provide specialty process piping. As it turned out, this "sideline" pursuit transformed Kinetics and forged lasting relationships with the leaders in dynamic new business segments — at a time when those leaders were still in their infancy.

In particular, a 1977 project catapulted Kinetics from a 15-person operation to over 100 employees virtually overnight. The subsequent decision by Kinetics to exclusively pursue mechanical contracting of high-purity process piping proved to be particularly well timed as the semiconductor industry blossomed, which in turn helped fuel a national expansion for the company. By leveraging its experience in the microelectronics and biopharmaceutical industries, Kinetics financed new equipment and quickly made inroads with new clients and industry categories. The company added AMD to its client list in 1978 and began work on Genentech's first major production facility the next year.

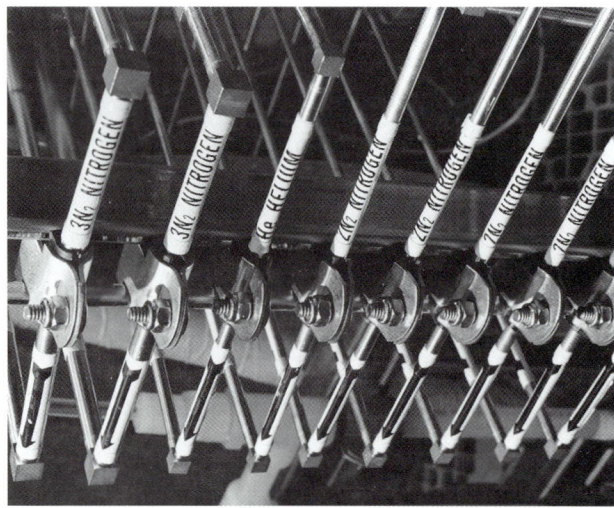

THE MOVE TO INTEGRATED SOLUTIONS

Early on, one of Kinetics' distinguishing characteristics was its understanding that mechanical contracting was just one facet of the related solutions that could help make its clients more successful. In a series of strategic acquisitions and alliances, Kinetics repeatedly added to the size and scope of its capabilities, achieving the full turnkey approach from a single source of accountability that now differentiates the company to customers around the world.

FOR SEMICONDUCTOR CUSTOMERS

For instance, in the semiconductor industry, there was a clear need for turnkey solutions as Kinetics professionals observed the difficulties involved in the traditional approach to designing or upgrading wafer fabrication plants. Fab plant owners often turned to Kinetics to provide high-purity piping for wafer fabrication plants, relying on the company's sophisticated knowledge of all issues related to process-piping installation for systems delivering the gases, chemicals and water used in semiconductor

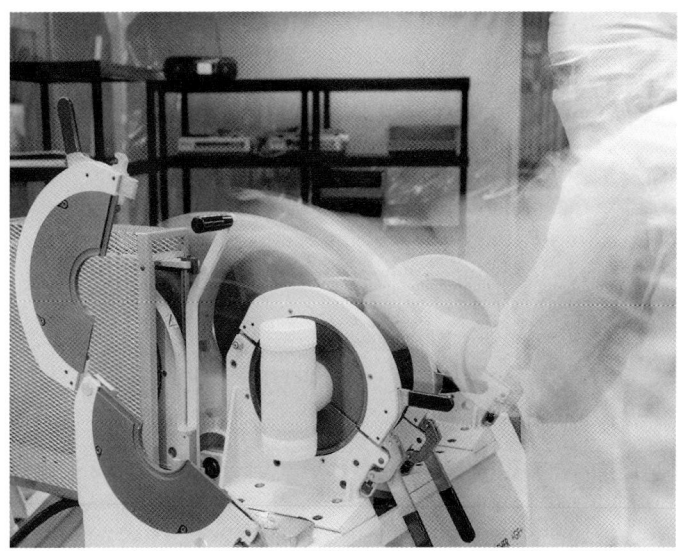

production. However, because projects typically brought together multiple vendors, plant owners faced the challenge of managing different companies through the various long and arduous project stages, which frequently led to inefficiencies and expensive production delays.

To remedy this situation, Kinetics acquired companies that broadened its design, manufacturing and installation services. Moving into integrated fluid systems added a technological solution to the firm's design and manufacturing capabilities, allowing Kinetics to provide manifolding and precise flow control of critical gases and chemicals from the source to the semiconductor manufacturing tool. Kinetics' Unit Instruments product line added a wide array of leading-edge mass flow control and fluid delivery instrumentation for inside the tool. Kinetics' Mega Systems acquisition provided critical chemical and slurry blending and distribution capabilities, while Kinetics' FTS Systems product line introduced high-quality, precision-controlled temperature solutions for fabrication and testing. Other transactions added solutions in ultrapure water systems, waste treatment, reclamation and re-use. Moreover, Kinetics tied together these critical process systems with process system engineering, ensuring reliable design and construction of the complete gas, chemical and water infrastructure, including hook-up of the water fabrication tools. Kinetics has performed projects for over 125 semiconductor clients, more than any other process systems provider in the world.

In another innovative arrangement, Kinetics partnered with Air Products and Chemicals to form TRiMEGA, which added to the company's ability to "surround the tool" by providing bulk and specialty gases and chemicals, as well as on-site operation services.

The net result is a company that offers semiconductor manufacturers expert in-house design resources, component and subsystem manufacturing, financing, and detailed program management to oversee and execute projects — complete with quality control and quality assurance — from conception to commissioning to continued on-site maintenance.

FOR BIOPHARMACEUTICAL CUSTOMERS

For the competitive biopharmaceutical industry, speed-to-market is a major economic issue, further complicated by the need to meet stringent purity standards by the federal Food and Drug Administration and other regulatory agencies.

In this arena, Kinetics took a similar turnkey approach, offering process system design, fabrication and a guaranteed validated installation. Besides process engineering and facility design, the company acquired ultrapure water and waste treatment systems, modular process systems from WHE Biosystems, industrial freeze drying systems from FTS Systems, and dry material blending and containment systems from Tote Systems.

To date, Kinetics has completed system installations for over 400 top biopharmaceutical clients, with over 75 percent of the company's work coming from repeat clients.

A GLOBAL LEADER

Today, Kinetics is both the largest global provider of integrated, high-purity process systems and the only company offering such a full array of design, manufacture and installation services. The company now has 25 construction offices and 23 manufacturing/service centers around the world to provide local service and support for its customers. Led by president and CEO David J. Shimmon, Kinetics is over a $1 billion organization of 8,300 employees with aggressive expansion plans to leverage its 25 years of experience and innovation into new markets around the world.

Integrated fluid delivery systems

Integrated process solutions

PeopleSoft

PEOPLESOFT, A WORLD LEADER IN E-BUSINESS SOFTWARE, IS HEADQUARTERED IN PLEASANTON, CALIFORNIA, about 30 minutes from the traffic-clogged arteries of the high-tech heartland. Technologically and culturally, however, PeopleSoft stands right at the core of the values that have kept Silicon Valley booming through three tumultuous decades.

Now a $1.4 billion company with 7,000 employees, 4,000 customers worldwide and a 50 percent share of the market for human resources (HR) software, PeopleSoft was founded in 1987 by entrepreneur Dave Duffield and

PeopleSoft corporate headquarters
FGI Print Management

his partner, Ken Morris. The pair was just slightly ahead of its time in recognizing that the ad hoc systems of mainframes and desktop PCs used for most corporate computing would soon be replaced by client-server networks.

Duffield founded PeopleSoft to develop applications for the emerging client-server environment. The goal was to make corporate databases available online and in real time to all employees, thereby providing a huge productivity boost for mid- to large-sized corporations.

HIGH-TECH VISION WITH A HEART

PeopleSoft was actually the third company founded by Duffield, who is himself a softer, more people-oriented version of the classic high-tech visionary. Duffield launched his first company soon after graduating from Cornell University. His idea was to custom-develop software for human resource management at colleges and universities. To promote this service, Duffield and friends visited college towns across the country. They would settle down for a couple of months whenever they found a likely institution, live in dorms while getting to know the people in the human resources department, and ultimately develop software that was custom-tailored to the university's needs. Moving slowly from east to west, and enjoying himself all the way, Duffield ultimately arrived in California where he successfully took on the challenge of UC Berkeley, one of the largest and most tumultuous universities in the nation.

Duffield's original goals for PeopleSoft had nothing to do with creating a billion-dollar corporation. He envisioned a firm of some 50 employees who would develop good software, satisfy customers, make a positive contribution and have fun doing it. By 1991, PeopleSoft had more than achieved these modest goals and was established in a profitable niche, serving about 100 happy customers of its human resource management programs. In those early days, company meetings were cozy, convivial affairs held at Duffield's home, a modest 25-acre spread in the exclusive Blackhawk development. The surge in client-server applications, however, was still gathering force, and Duffield realized that PeopleSoft was perfectly positioned to ride further than he had ever anticipated.

From 1991 to 1994 the company branched out into a series of new applications, including software suites for financial management, manufacturing processes and distribution networks. Meanwhile, PeopleSoft's HR programs were establishing themselves as the worldwide favorite. Gradually, the company built a set of generalized applications designed to facilitate the operations of almost any business. The next step was to specialize, optimize and integrate these applications for particular industries. PeopleSoft decided to focus on a few new fields, such as financial services and health care, while retaining its traditional base in higher education and federal government services.

FROM CLIENT-SERVER TO THE INTERNET

In 1994 the surge in client-server applications slowed as the Internet began to define its role as a powerful and potentially unlimited alternative to local or wide-area networks. For PeopleSoft, the emergence of the Internet was something of a blessing in disguise. On the downside, the company wasn't driving new technology as it had been in the early client-server days. On the upside, the PeopleSoft vision of automating and integrating enterprise applications was eminently adaptable to the era of e-commerce.

Since the mid-90s, the emphasis at PeopleSoft has been on Internet enablement of existing software and the integration of ERP (Enterprise Resource Planning, the company's traditional strength) with end-to-end solutions for selling and buying over the Web. PeopleSoft 8, the company's latest release, features 100 percent Internet enablement. An integrated suite of applications and tools for global e-commerce, PeopleSoft 8 includes modules for the management of enterprise performance, human resources, financials, service supply chains, materials and manufacturing.

Prominently featured in PeopleSoft 8 are two new e-commerce solutions called eStore and eProcurement. eStore provides a Web storefront designed for trouble-free integration with a company's e-business backbone. Many companies have begun taking orders over the Internet only to find that they have no online way of accessing such vital data as available inventory, order status or customer profiles. eStore facilitates integration between order capture over the Internet and order management, fulfillment and analysis. Users are finding that the payoffs include lower transaction costs, improved efficiency, better analysis and greater profitability.

PeopleSoft eProcurement helps companies streamline the procurement of indirect goods and services (such as office, electrical and audiovisual supplies). In many corporations, the tedious manual processes now used in procurement lead to inefficiencies, higher prices, extended cycle times and little or no analysis of supplier performance. The benefits of forging automated ties to key suppliers can be huge. Since many organizations spend up to 60 percent of their revenues on indirect purchasing, a 5 percent reduction in costs could lead to a 28 percent boost on the bottom line.

Business-to-business (B2B) e-commerce over the Web is expected to top $1 trillion by 2003 with another $109 billion tacked on by business-to-consumer (B2C) sales. With applications such as eStore and eProcurement, plus 100 percent Internet enablement of its ERP software, PeopleSoft sees abundant opportunities for short- and long-term growth.

Will continuing success force PeopleSoft away from its fun-loving, humanistic roots? Insiders say not. In about 10 years, the company has grown from a start-up to a global corporation with billion-dollar revenues and thousands of employees. The company's quarterly meetings now have to be held in cavernous

© 1999 Lonnie Duka

buildings at the Alameda County fairgrounds — the only nearby venue large enough to accommodate everyone. But the core PeopleSoft values remain intact. Duffield has begun to distance himself from day-to-day operations, but there is still an emphasis on work-life balance as well as a non-hierarchical structure that values each employee, encourages access to top executives and resists the kind of stratification that stifles innovation (not to mention individuality) at many large firms.

Fun on the job remains one of the company's core values, and employees must be taking the pleasure principle to heart because *Fortune* ranked PeopleSoft No. 6 on last year's list of the "100 Best Companies To Work For" in America. Competition to hire engineering talent is as vital to the health of a Silicon Valley enterprise as competition to capture markets. PeopleSoft believes that its unique corporate culture and accessible location will make it a magnet to the best and brightest for many years to come.

© 1999 Lonnie Duka

SEMI

TODAY THE SEMICONDUCTOR EQUIPMENT, MATERIALS AND SERVICES INDUSTRY DRIVES THE SEMICONDUCTOR manufacturing industry, which in turn supplies the chips that allow electronics manufacturers to create new computer, communications and Internet products. But it wasn't always so.

In the industry's early years, the chip manufacturers developed most of their own processing equipment, ordering the parts and materials they required from equipment and materials companies who were treated as "vendors" of silicon wafers, chemicals, gases and production

SEMICON Singapore is one of many expositions SEMI produces around the world.

equipment but seldom were involved in its use. Whenever the chip manufacturers needed new processing equipment or materials, they would put out a specification defining their needs, and the equipment and materials vendors would submit a bid and hope for the best.

It is a different world today. The semiconductor devices of today are so complex that progress to new generations cannot be made without advances in equipment and materials, and thus companies in that market have become valued "partners" in the manufacturing process.

As their importance to the large integrated circuits (chips) producers increased, equipment and materials companies needed a venue for showing their latest products to multiple customers. Whereas in the past they had been relegated to the back aisles of the larger components trade show, in 1970 a group of them got together and decided to start a show of their own, focusing on every aspect of semiconductor production. The event, known as SEMICON, became the first step in defining what is now one of the world's fastest-growing industries. The founders, however, were interested in selling products, not in managing trade shows. Thus, they decided to form an independent trade organization of their own.

Dedicated exclusively to its industry segment, the new organization formed in 1970 called itself the Semiconductor Equipment and Materials Institute (SEMI). The organization's first trade show, SEMICON, was held in 1971 at the San Mateo County Fairgrounds. At the time, a trade show serving a unique segment of the semiconductor industry was considered an entrepreneurial undertaking. But it proved a success, and by year 2000, SEMICON was regularly hosting two U.S. shows and similar events in Europe, Korea, Taiwan, China, Japan and Singapore.

Today SEMI, renamed Semiconductor Equipment and Materials International to reflect its global interests, is a worldwide, non-profit trade association that represents the semiconductor and flat-panel-display equipment and materials industries. The organization is represented in all significant semiconductor manufacturing regions around the world, and maintains offices in Asia and the Pacific, Europe, Japan, North America and Russia.

From the earliest days, SEMI members have played a significant role in creating better semiconductors. SEMI members, working with semiconductor manufacturers, have enabled more advanced electronics. Over the years, SEMI has always recognized the inventors and entrepreneurs in the industry. Starting in 1977, SEMI began its recognition program by honoring the inventors of the transistor: William Shockley, Walter Brattain and John Bardeen. Without the invention of the transistor, life as it exists today would be quite different. SEMI has always prided itself on recognizing the inventors and entrepreneurs that have made the industry great. Today, SEMI has a global program recognizing the significant contributions made by entrepreneurs in semiconductor-related technology.

As time went on, the organization was also quick to recognize the need for market data, technical information, advocacy on public policy issues, and industry manufacturing standards. Also high on its early agenda was helping member companies to expand in overseas markets.

In its early attempts to increase industry sales overseas and fight trade barriers, SEMI took some of its first steps toward making the semiconductor industry truly global. By 2000, approximately 45 percent of SEMI's membership consisted of non-U.S. companies, whereas there were no non-U.S. members in 1970.

Between 1997 and 2000, SEMI's corporate membership grew by 700 members; many were small companies that wanted to learn more about the industry and/or gain industry recognition. And by the year 2000, when computer-based automation had become essential to chip production, about 15 percent of SEMI's membership was providing software-related services to the semiconductor industry. This industry segment, which did not even exist 20 years ago, is now one of the fastest-growing businesses in the sector.

One of the unique challenges facing this organization has been to find a way to operate both globally and regionally. As a result, SEMI has learned to treat matters of policy and practices globally, and issues of implementation and operations locally (and in local languages). "Being global ensures that all members can play on level terrain. Acting locally helps get things done faster and better serves the needs of our members in those markets," says Stanley T. Myers, President of SEMI.

Many of SEMI's leaders have been or are members of the organization's board of directors. Emeritus board members are those individuals who have made outstanding contributions to SEMI during at least 11 years on the board. Three noteworthy directors emeritus have been Shigeo Takayama, President of Hakuto Co., Ltd. (Japan); James C. Morgan, Chairman and CEO, Applied Materials, Inc. (USA); and Ken Levy, Chairman of the Board, KLA-Tencor (USA).

Around 1998, SEMI began setting up regional advisory boards. These boards are typically made up of local representatives from academia, semiconductor chip companies and SEMI members. Their *raison d'etre* is to provide an opportunity for local ideas to reach the functional committees and the international boards.

Today SEMI regularly facilitates the exchange of information and promulgates worldwide industry standards by sponsoring conferences, training courses, trade missions and expositions throughout the world. It is constantly generating new information products, videos, newsletters and Internet resources to provide timely business, marketing and technical information.

The organization is well known in the United States for its public policy efforts in the areas of export controls, government support for the industry and reduction of tariff and non-tariff barriers. SEMI maintains government relations for the purpose of advocating free and open-market access worldwide and addressing member concerns on import/export controls, trade policy and tax issues.

Stanley T. Myers, President, Semiconductor Equipment and Materials International (SEMI)

In celebration of its 30th anniversary in July of 2000, SEMI pledged $1 million to Silicon Valley organizations helping prepare students for high-tech careers. This decision came as both a response to the White House's call for efforts to narrow the digital divide, and to the recognition that the semiconductor industry will continue to need innovative people to maintain its current rate of growth. Also in 2000, SEMI moved from its headquarters in Mountain View to a new facility in San Jose.

To meet the challenges of 21st century, SEMI plans to continue sponsoring conferences, symposia, standards developments meetings, expositions and other activities that will add value to the technology of the organization's members. But above all, explains SEMI's Myers: "SEMI will aim to be the organization of choice that helps its members on a pre-competitive basis collectively attack business technical and operational issues."

SEMICON West 77: (left to right) William Shockley, Charles C. Harwood, Walter H. Brattain, Sheldon Weinig, John Bardeen, Patrick E. Haggerty

Silicon Image

THE WORLD IS "GOING DIGITAL." AND, WHILE SUNNYVALE-BASED SILICON IMAGE MAY NOT BE A HOUSEHOLD NAME in Silicon Valley, it is this company that in 1999 pioneered the development of worldwide industry standards for low-cost, high-speed digital data transmissions. Adoption of this standard has already had a revolutionary effect on the digital world.

Silicon Image, founded in 1995, designs, develops and markets semiconductor solutions for high-speed digital communications. Its technology is designed for applications that require cost-effective, high-bandwidth integrated solutions for data transmission, such as the local interconnect between host systems and digital displays, high-speed networking and data storage. To a consumer, this results in two chief benefits — swift transmission of data and sharper, clearer video images for computers, consumer electronic devices and displays.

FOUNDED ON A DREAM

Dr. David D. Lee, Silicon Image chairman, president, CEO and the company's principal founder, had a dream to found his own company while working on a doctorate degree at UC Berkeley and building a new high-performance computer with a colleague, Dr. D.K. Jeong. Following graduate school, Lee spent six years on the research staff at Xerox Palo Alto Research Center (PARC) and was later a principal investigator at Sun Labs before founding Silicon Image. Jeong worked for Texas Instruments before becoming a professor at Seoul National University (SNU). The ties between the men have remained strong, and today SNU students gain experience in their graduate program through a cooperative working agreement with Silicon Image.

PRODUCTS AND CAPABILITIES

Silicon Image's product line, the PanelLink® digital family, is based on the company's proprietary technology that provides high bandwidth for digital displays, connecting host systems (transmitters), displays (discrete receivers) and controllers that integrate PanelLink receiver technology with other functionality to enable intelligent displays for the mass market. PanelLink is an industry leader both in terms of technology and market share.

Silicon Image currently has two target markets — host systems and displays. Host systems include desktop and notebook PCs, set-top boxes and DVD players. Displays include liquid crystal display (LCD) monitors, flat panel displays, digital cathode ray tubes (CRTs), high definition TV (HDTV) displays, projectors and notebook displays.

PanelLink technology transmits data over three high-speed serial channels for an aggregate bandwidth of approximately five gigabits per second. The technology supports such speeds over inexpensive twisted-pair copper wire at distances of up to 10 meters and permits direct coupling with fiber-optic interconnect modules for longer distance data transmission. The result is an all-digital video solution that replaces flicker, fuzziness and color variation with a sharp, crisp, crystal-clear video image at a low cost. It is this combination of digital quality at analog prices that makes this technology appealing for mainstream business and consumer applications.

Silicon Image has shipped over 10 million units within three years. Its products are incorporated in host systems and displays sold by leading manufacturers such as Apple, Compaq, Dell, Fujitsu, Gateway, HP, Hitachi, IBM, LG, Matrox, NEC, Samsung, Sharp, Sony, Toshiba and ViewSonic.

PIONEER IN DIGITAL TECHNOLOGY

Two prominent trends in the electronics industry have created opportunities for Silicon Image to go where no

David Lee, founder and CEO

semiconductor solutions provider has gone before. The first is the increasing demand for bandwidth — the amount of data that can be transmitted across a medium in a given period of time, often measured in megabits per second. The second is the transition of electronic systems from analog to digital communications, which has made it easier to reliably store, transmit and manipulate information.

The blessings of a digital world are almost immeasurable. For example, many features such as messaging, paging and security have become more feasible as wireless phones have migrated from analog to digital format. Likewise, significant benefits including clear, sharper images can be achieved by replacing analog displays with digital displays. But, to enable the transition from analog to digital, displays require digital communications with the host system. And until recently, there was no commercially viable standard that addressed the challenges of enabling digital communications between host systems and video displays. A chief drawback for such a standard was the substantial technical challenge of developing a cost-effective, high-bandwidth solution capable of transmitting data at the multi-gigabit rates needed to link the host system to a high-resolution display.

Recognizing the need for a cost-effective, high-bandwidth solution, Silicon Image developed its digital interconnect technology and began shipping semiconductor products for digital displays in 1997. Then, to provide a worldwide, open specification for an all-digital display solution, Silicon Image, together with Intel, Compaq, IBM, Hewlett-Packard, NEC and Fujitsu, formed the Digital Display Working Group (DDWG) to define such a specification. Using Silicon Image's technology as its foundation, the DDWG published the Digital Visual Interface (DVI) specification in April 1999 to define a high-speed universal data communication link between host systems and displays.

INDUSTRY BENEFITS OF DVI SPECIFICATION

The creation of the DVI specification was truly a benchmark in the history of the semiconductor industry. Today, more than 100 companies, including systems manufacturers, graphics semiconductor companies and monitor manufacturers, are participants in the DDWG, and many are developing hardware and software products designed to be compliant with the DVI specification. The DDWG's release of the DVI specification — royalty-free, to encourage the entire industry to benefit from it — has accelerated the shift to digital display technology. DisplaySearch, a market research firm, projects that the number of desktop LCD monitors shipped annually will grow from 4 million in 1999 to 30 million in 2005, and that the digital interface will rapidly gain market share over analog for desktop LCDs, from 30 percent in 1999 to over 90 percent in 2003.

Another consequence of the DVI specification is seen in the CRT market. The top five desktop CRT manufacturers are developing digital CRTs that are compliant with the DVI specification. These companies are looking for integrated semiconductor solutions that combine high-speed digital communications technology with the functionality required to enable intelligent displays for the mass market.

Silicon Image's challenges in this highly competitive industry are to make technology better and to bring more innovative products to an expanding market. Low-cost, high-bandwidth transmission capabilities and Silicon Image's core technology provide applicability for multiple mass markets such as networking and data storage. The visionary zeal that made Silicon Image a pioneer in creating DVI standards still drives the company as it works on new solutions for businesses, from migrating flat panel displays to digital format to the massive opportunities for viewing high-definition content on digital TVs for consumers.

Silicon Image headquarters in Sunnyvale

Digital displays using PanelLink technology for highest quality digital image

Winbond

Chairman Arthur Chiao and President C.C. Chang's vision has enabled Winbond to become a global player in the highly competitive semiconductor market.

BACK IN 1987, A SMALL GROUP OF TAIWAN'S TOP ENGINEERS LAUNCHED A SEMICONDUCTOR STARTUP CALLED Winbond. Their goal was to make the new company a key player in Taiwan's budding IC industry. With timing and talent on its side, Winbond rapidly outgrew its founders' expectations. Today, Winbond integrated circuits (ICs) are not only the dominant brand in Taiwan, but they are also hard at work in the vast majority of American households, embedded in such indispensable products as cell phones, digital answering machines, VCRs, PCs and talking toys.

Winbond achieved its initial goal of becoming a major IC supplier to Taiwan-based PC subcontractors and consumer electronics companies within its first two years. By the early 1990s, Winbond had emerged on the international scene as one of world's top three suppliers of ICs for telephone dialers, a position it continues to hold today. By 1995 the company had achieved $600 million in revenues and realigned its strategy for success in the Internet era. Realizing that the Web would create an explosion in new applications, Winbond began developing IC solutions for Video CD players, DVDs, video phones, MPEG decoder chips, Web browsers and other emerging consumer entertainment appliances. Today, Winbond is the worldwide leader in ICs for speech recognition and synthesis.

The company is also a major force in high-density DRAMs (Dynamic Random Access Memories), which are essential to the performance of today's personal computers. Winbond celebrated the millennium by releasing a 256Mb DRAM fabricated with a state-of-the-art, 0.175-micron process. As part of a strategic alliance with Toshiba of Japan, Winbond will soon manufacture DRAMs with an even more sophisticated 0.13-micron process. DRAMs and flash memories account for about 30 percent of Winbond revenues. The company's other semiconductor lines include data communications, microcontrollers, multimedia, telephony and graphics ICs.

Winbond's state-of-the-art manufacturing fabs will produce more than 30,000 wafers a month in 2000.

Winbond's four facilities for semiconductor fabrication in Taiwan have a combined capacity of more than 100,000 wafers per month. High-tech journalists from around the world make a point of touring the Winbond fabs, which are among the most advanced in the industry. These resources have made Winbond a trusted supplier to such high-tech industry giants as IBM, LSI Logic, Toshiba, NEC, Sony, Dell, Seagate and Compaq.

The company now has 4,000 employees worldwide. It has always recruited engineering talent from around the world, and offers exciting career opportunities in Taiwan, Israel, the United States and Europe. In terms of providing a cosmopolitan and professional working environment, Winbond is on par with any U.S.-based high-tech company.

INVESTMENT IN R&D

Investment in R&D has been a top priority at Winbond since the very beginning. The company now holds more than 500 worldwide patents, has an additional 700 patents pending, and ranked among the top 300 companies in U.S.-patent applications in 1998. Winbond earned the Gold Medal in Taiwan's national invention competition in 1999. In an international survey of electronic component suppliers in 1999, Winbond won first place in four categories: innovation, product quality, service quality and best performance.

At Home in Silicon Valley

Winbond came to Silicon Valley in 1990, establishing a wholly owned subsidiary to concentrate on engineering, sales and marketing in North America. WECA (Winbond Electronics Corporation of America) was also charged with acquiring synergistic technologies and companies that complement the parent corporation's goals and market focus. To this end, Winbond acquired Symphony Labs (chipsets for PCs) and Winbic (cache RAM for high-speed computing) in the mid-90s. In 1998, Winbond acquired ISD (Information Storage Devices), a classic Silicon Valley startup that had enjoyed rapid growth as a premier provider of voice solutions in silicon.

ISD was founded in 1987 to develop voice solutions on the semiconductor level. The company was the brainchild of two Silicon Valley veterans who had roots stretching all the way back to the early days of Intel. ISD's founders believed that speech was the key to streamlining and humanizing the man-machine interface. They also believed that the voice activation solutions on the market at that time would never unleash the true potential of speech because they were based either on cumbersome software or on chipsets demanding massive amounts of memory.

Starting with a core group of about a dozen people and a technique that vastly expanded the amount of voice information that could be crammed into a single memory cell, ISD found ways to drastically reduce the cost of speech storage in silicon. Soon the company was able to offer chips that provided up to 10 seconds of speech storage at about 50 cents per chip, a cost that made the devices ideal for consumer products. In 1994, Hallmark bought nearly 5 million ISD speech chips for greeting cards, and that was just an initial order. Motorola also began using ISD chips for voice memos in 1994.

With large orders came rapid growth. ISD went public in 1995, before the boom in tech stocks that made almost any Silicon Valley IPO a hot prospect. But ISD had all the fundamentals for success, including good people, products, profits and experienced management. ISD was voted one of the fastest-growing companies in the Valley in 1995. By 1996, ISD had about 150 people on board and was a recognized leader in the technology and marketing of chip-based speech solutions.

In 1997, however, ISD hit a plateau. The leaders had exciting ideas and felt that they were just beginning to tap the potential of voice solutions in silicon; but to move into new product areas, they needed more financial resources, access to large foundries and Intellectual Property rights to microprocessor cores and other chip technologies associated with speech solutions. At the same time, Winbond recognized that its corporate goals of international expansion and leadership in speech ICs would be well served by the acquisition of ISD's technologies, expertise and marketing network.

The acquisition helped extend Winbond's leadership in voice ICs. It also gave the ISD technologists the resources they needed to make rapid progress in creating new sophisticated voice ICs that will help users gain the full benefits from increasingly complex and miniaturized products in fields such as communications, home electronics and automotive sound systems.

The strong economies in America and Europe, combined with signs of recovery in Asia and a significant drop in costs for capital equipment in the semiconductor industry, give Winbond compelling reasons for optimism about its future. With its aggressive position in DRAM process technologies and its leadership in voice ICs, Winbond plans to contribute IC solutions to a host of hot products in coming years, including the one that just might make it a household name: a cell phone equipped with a voice recognition chip that allows drivers to keep their eyes on the road and their hands on the steering wheel.

Winbond's "Winning Solutions" philosophy enables customers to leverage more than 400 IC products across a broad range of applications.

BIBLIOGRAPHY

The listings below helped provide either information in the research of this book, or were purely inspirational in adding to the texture of *Sand Dreams & Silicon Orchards*.

REPORTS:

Hylkema, Mark G. 1995 Archaeological Investigations at *The Third Location of Mission Santa Clara de Asis: The Murguia Mission, 1781-1818.* Report prepared for the California Department of Transportation, District 4, Oakland. Reprinted through Coyote Press, Salinas, CA.

BOOKS:

Baxter, Don. *Missions of California.* Pacific Gas & Electric Co., 1970.

Bean, Lowell John. *The Ohlone Past & Present.* Ballena Press, 1994.

Beilharz, Edwin and Donald DeMers Jr. *San Jose: California's first city.* San Jose Chamber of Commerce, 1980.

Brunelle, Mark. *Junipero Serra.* Dobronte Publications, 1984.

Butler, Phyllis Filiberti. *Old Santa Clara Valley.* Wide World Publishing/Tetra, 1991.

Camarillo, Albert. *Chicanos in California.* Boyd & Fraser Pub. Co., 1984.

Crowley, David and Paul Heyer. *Communication in History.* Longman, 1993.

Dimmick, Donna. *Union Cemetery Redwood City, CA.* Historic Union Cemetery Association, 1999 (pastlinks.com/huca).

Farrell, Harry. *San Jose — and other famous places.* History San José, 1983.

Fleming, Paula Richardson and Judith Lusky. *The North American Indians in Photographs.* Barnes & Noble Books, 1986.

Gullard, Pamela and Nancy Lund. *History of Palo Alto.* Scottwall Associates, 1989.

Jacobson, Yvonne. *Passing Farms: Enduring Values.* California History Center, De Anza College, http://wwwdeanza.fhda.edu/CalifHistory/CalifHistory.html., 1984.

Loomis, Patricia. *A Walk Through the Past.* Argonauts Historical Society of San Jose, 1998.

Lyman, George. *The Ralston Ring.* Charles Scribner's Sons, 1936.

Mirrielees, Edith R. *Stanford, the story of a university.* G.P. Putnam's Sons, 1959.

Norton, Henry K. *Story of California.* A.C. McClurg & Co, 1913.

Olmert, Michael. *The Smithsonian Book of Books.* Wings Books, 1992.

Osio, Antonio Maria. *The History of Alta California.* The University of Wisconsin Press, 1996.

Payne, Stephen. *Santa Clara County: Harvest & Change.* Windsor Publications, 1987.

Pierce, Marjorie. *San Jose and its Cathedral.* Western Tanager Press, 1990.

Rambo, Ralph. *Remember When, A boy's-eye view of an old valley.* Rosicrucian Press, 1965.

Reflections of the Past, Heritage Media, 1996.

Reinstedt, Randall A. *Portraits of the Past.* Monterey Savings, 1979.

Schneider, Jimmie. *Quicksilver: the complete history of Santa Clara County's New Almaden Mine.* Published by Zella Schneider, 1992.

Watkins, T.H. *California, an illustrated history.* American West Publishing Company, 1973.

White, Stewert Edward. *The Story of California.* Halcyon House, 1940.

William, James C. *The Rise of Silicon Valley.* California History Center, De Anza College.

NEWSPAPERS:

Selected articles from the *Daily Grind* (1993), *San Jose Mercury-Herald* (1906) and *San Jose Mercury News* (1992-1996).

MAGAZINES:

Selected articles from *Newsday* (1999), *Sports Illustrated* (1991) and *USA Today* (1996).

INTERNET SITES:

http://www.sfmuseum.org
"Earthquake and Fire: San Francisco in Ruins." *Call-Chronicle Examiner* article, Museum of the City of San Francisco.

http://ac.acusd.edu/History/recording/notes.html
Ampex

http://encarta.msn.com/find/Concise.asp?ti=00493000
Apple

http://www.apple-history.com/history.html
Apple

http://gi.grolier.com/wwii/wwii_1.html
Department of Military Art, United States Military Academy

http://www.scu.edu/SCU/Programs/Diversity
Diversity

http://www.internetcenter.state.mn.us/Itn-083.htm
ENIAC

http://www.islandnet.com/~kpolsson/comphist.htm
History of the computer

http://ei.cs.vt.edu/~history/index.html
History of the computer

http://history.nasa.gov/brief.html
History of NASA

http://www.fas.org/cp/pong_fas.htm
History of Pong

http://encarta.msn.com/find/Concise.asp?ti=00493000
Jobs, Steve

http://ei.cs.vt.edu/~history/Jobs.html
Jobs, Steve

http://www.dreamscape.com/joanstem/toc.html
Jobs, Wozniak, Bushnell, McCallum

http://www.comet.arc.nasa.gov/jf/mfa/history.html
Moffett Field History

http://businesstech.com/feature/btinxight9703.html
PARC

http://stn2.net/transistor/album1/shockley
Shockley, William

http://www.webstationone.com/fecha/shockley.htm
Shockley, William

http://www.nobel.se/laureates/physics-1956-1-bio.html
Shockley, William

http://vacuumtube.com
The history and future of the vacuum tube

http://www.service.com/PAW/morgue/news/1995_Jan_4.CRE-ATR44.html
Varian, Russell

http://www.varian.com/corp/history.htm#Vacuum
Varian, Russell

http://www.varian.com/corp/vintage/vv1e.html
Varian, Russell

http://www.cyberhikes.com/HCRPINFO.HTM
Varian, Russell & Castle Rock

http://www.pbs.org/battlefieldvietnam/history/index.html
Vietnam

http://encarta.msn.com/find/concise.asp?z=1&pg=2&ti=03512000
Wozniak, Steve

http://www.execpc.com/~shepler/wozniak.html
Wozniak, Steve

http://www.posi-tone.com/BOB/html
http://members.aol.com/alhidell/paranoia/22Article.htm
Zappa, Frank

http://www.mediahistory.com/time/1950s.html
1950 technology timeline

http://ericir.syr.edu/Virtual/Lessons/crossroads/
Unit IX.   Boom and Bust, 1921-1933
Unit X.   The Age of Franklin D. Roosevelt, 1933-1945
Unit XI.   Leader of the Free World, 1945-1975
Unit XII.   A Nation in Quandary, 1975

VIDEO:
Nerds 2.0.1, Bob Cringely, A Brief History of the Internet, PBS Home Video, 1998.

Triumph of the Nerds, Bob Cringely, Ambrose Video Publishing, Inc. 1996.

AUDIO:
Interview with Russell and Sigurd Varian in a 1958 taped interview housed in Stanford University's archives.

INDEX

Academy of Motion Picture Arts and Sciences	96
Agnews Sanitarium	60, 61
"Albert of Belgium"	66
Alta California	18
Alviso Station	53
American Ambulance Corps	66
Anderson Dam	123
Andrew, Dane	36
Andy Caps Tavern	102
Año Nuevo	20
Apple	102, 103, 104
Artemis Ventures	130
AT&T Bell Telephone Laboratory	77, 78, 81
Atari	103
Autodesk	125
Baird, Harold	119
Baird, Johne	117
Baldwin, Ella	64
Baran, Paul	96
Bardeen, John	77
Bartz, Carol	125, 127
Bay Area	14, 31, 60, 93, 115, 137
Bayshore Freeway	70
Beckman Instruments	79
Beddo, Carol	58
Bezos, Jeff	123
Blade Runner	103
Blake, William	49
Blank, Julius	79
Bocks, Charles	55
Booth, John Wilkes	37
Bow, Clara	55
Brattain, Walter	77
Bromfield, John	64
Brookhaven National Laboratory (BNL)	101
Brown, Gov. Jerry	93
Burlingame Country Club	64
Burnet, Dana	66
Bushnell, Nolan	100
Business Week Magazine	124
Cabrillo, Capt. Juan Rodriguez	16
California Prune and Apricot Growers Association	39
CalTrans	136
Camp Kearney, San Diego	69
Carlos, John	94
Carolan, Francis	63-65
Carquinez Straights	21
Castle Rock	74
Central Pacific Railroad Company	41
Chambers, John	131
Charles III	18
Chavez, Cesar	93, 94
Chinese Exclusion Act	53
Chinese Historical and Cultural Project	53
Chip Garden	114
Chronicle	65
Chuck E. Cheese Pizza Time Theater	102
Chynoweth, Mary Hayes	37-39
Chynoweth, Thomas	39
Cisco Systems	115, 117
Civil Liberties Act of 1988	99
Civil War	36
Civilian Conservation Corps (CCC)	69
Claremont Hotel	65
Clark, Kent	135
Clarke, Arthur C.	78, 132
Cleveland, President Grover	45
Clinton, President William	98
Cody, William F. "Buffalo Bill"	41
Cold War	76, 96
Colorado River	16
Comaford, Christine	129
Community Service Organization (CSO)	93
Congressional Asian Pacific Caucus and Research Group	99
Coolidge, Charles Allerton	43
Cornell University	64
Coyote Creek	123
Crespi, Father Juan	19
Crocker, Charles	40
Crocker, Ethel	65
Crossways Farm	64
Cullers, Kent	136
Cupertino	52, 103, 115
Daly City	116
Dandini, Countess Lillian Virginia Remillard	66
Dartmouth College	96
Defense Department Advanced Research Projects Agency (DDARPA)	96
Del Monte Foods	58
DeNeve, Felipe	21
Diaz, Capt. Melchior	16
Dick, Philip K.	103
Dinosaur World	112
Direct Access Storage Device (DASD)	78
Drake, Sir Francis	16
Drakes Bay	16
East Bay	60, 116
East Palo Alto	137
East San Jose	93
Eckert, John Presper	77
Eden, Barbara	95
Edex	82
Edwards, Harry	94
Eitel-McCullough	71
Elwell, Cyril	67
ENIAC	77
Fages, Pedro	21
Fair Labor Standards Act (FLSA)	68
Fairchild Camera and Instruments	80
Fairchild Semiconductor	80, 81, 82
Fairmont Hotel	66
Fantasia	77
Farnsworth, Philo T.	72, 73
Farrell, Harry	92

Federal Arts Project	69
Federal Deposit Insurance Corporation (FDIC)	68
Federal Emergency Relief Association (FERA)	68
Federal Savings and Loan Insurance Corporation (FSLIC)	68
Federal Telegraph	67
Filoli	64
Folsom, Mary	38
Food Machinery Corporation	75
Forbes	123
Ford Philco	78
Fort Smith, Arkansas	98
Fountain Alley District	113
Fransanquito Creek	20
Fremont Camp	57
Frontier Village	112
Frost, David	93
Galvez, General José de	17-19
Garcia, Lorie	23
Garfield, Dr. Sidney	69
Garr, Dixie	131
Gates, Bill	129
General Electric	78
Gibbons, James	80
Gold Rush	21, 31, 34, 52, 114
Golden Gate Bridge	70
Good Government League	38
Grand Coulee Dam	69
Graphical User Interface (GUI)	97
Great Depression	68, 93
Grinich, Victor	79
Grove, Andy	85
Guadalupe Bridge	97
Guadalupe River	21
Half Moon Bay	115
Hambricht, Bill	104
Hammett, Dashiell	75
Hansen, William	73
Hatch, Harold	69
Hayes Conference Center	39
Hayes Estate	37, 113
Hayes, Daphne	38
Hayes, Elystus L.	38
Hayes, Everis Anson	38
Hayes, Janet Gray	99
Hayes, Jay Orley	38
Hayes, President Rutherford B.	38
Heart Mountain, Wyoming	99
Heinlein, John	53
Heinlenville	52, 53
Hewlett, Bill	76, 133
Hewlett-Packard	57, 103
High Rise Farm	120
Higinbotham, William	101
Hi-Life Hotel	114
Hiller Aviation Museum	134
Hiller, Stanley Jr.	134
Hillsborough	63
Hitler, Adolph	71
Hoerni, Jean	79
Hollings, Lawence	112
Hoover Bonneville Dam	69
Hoover, President Herbert	68, 70
Hopkins, Mark	40
House of Blues	113
Huntington, Collis	40
Hylkema, Mark	19, 34
IBM	78, 96
Inigo, Lope	22
Jacobson, Yvonne Olson	56, 57
Japan	74-75
Japantown	53, 99
Jobs, Steve	102, 103
Johnston, Adam	25
Jones, Earle	120, 121
Jones, Nicholas	120, 121
Jordan, Dr. David Starr	43, 67
Joshua Hendy Iron Works	62
Kaiser Shipyards	69
Kaiser, Henry J.	29, 69
Kawasaki, Guy	103
Kemeny, John	96
Kennedy, President John F.	89, 93, 96
Kennerson, Grace	54, 55
Kesey, Ken	97
King, Martin Luther Jr.	93
Kleiner Perkins Caufield & Byer	83
Kleiner, Eugene	79, 80, 81
Kurtz, Thomas	96
La Pérouse Expedition	14
LaFayette Fund	66
Lang, Marshall	74
Las Sergas de Esplandian	15
Last, Jay	79
Lawrence, Ernest	68
Liberty Ships	69
Lincoln, Mary Todd	37
Lincoln, President Abraham	36, 37, 64
Lincoln, Robert Todd	64
Linvill, John	80
Litton Electronics	71
Litton, Charles	71, 77
Lohse, William	129
Loma Prieta earthquake	60
Los Angeles Aqueduct	69
Lugosi, Bela	55
Malone, Michael S.	114
Manhattan Project	75, 101
Marimba	132
Markkula, Mike	103
Martini Creek	20
Mauchly, John W.	77
Menlo Park	127
Merck, Hannah Louise	56
Mercury-Herald	61

Mexican Heritage Garden	114
Middlebrook Gardens	113
Middlebrook, Alrie	113
Miller, Josephine	54, 55
Mineta Insurance Agency	98
Mineta, Norman	98
Mission Dolores	11
Mission Santa Clara	21, 22, 23
MIT	42, 68
Moffett Field	22, 70, 71
Moffett, William	70
Moniz Vineyard	123
Moniz, Stanley	122
Montalvo, Garci Rodríguez Ordonez de	15
Monterey Bay	14, 16
Moore, Gordon	79
Morgan Hill	122
Morissette, Jerry	136, 137, 138
Mussolini, Benito	71
National Academy of Sciences	79
National Farm Workers Association (NFWA)	93
Neumann, Janos	127
New Almaden Quicksilver Mining Company	34, 35
New Almaden	21
Newmark, Craig	128
Ng Shing Gung	52
Nixon, President Richard	91, 92
Norman, Peter	94, 95
Notre Dame High School	55
Noyce, Bob	79
Olson, Carl Johan	56
Olson, Rose Zamar	56
Olson, Ruel Charles	56
Olympic Games	95
Oppenheimer, Robert	75
Orwell, George	103
Packard, David	76, 114, 133
Page, George	38
Pajaro Indians	20
Pajaro River	19
Palo Alto Research Center (PARC)	97
Palou, Father Francisco	21
Panama-Pacific International Exposition	62, 63, 66
Park, James	97
Parkhurst, Mathias	33
Pasteur, Louis	109
Paul G. Allen Center for Integrated Systems	81
PC Magazine	129
Pearl Harbor	74
Permanente Cement Corporation	69
Polese, Kim	132
Polk, Willis	64
Pony Express	41
Porter, Bruce	64
Portola Valley	20
Portola, Capt. Gaspar de	18, 21
Presidio Monterey	21
Public Works Administration (PWA)	69
Pueblo San Jose	21, 34
Pullman Free School of Manual Training	64
Pullman, George	36, 63
Pullman-Carolan, Harriett	37, 63-66
Queen Elizabeth I	16
Ralston, William Chapman	63
RAMAC	78
RAND Corporation	96
Redwood City	32, 33
Redwood Ravine	33
Remillard, Pierre	66
Richards, Gary	136
Richards, Gilbert	33
Ridder, Joe	92
Roberts, Mike	112
Roberts, Sheldon C.	
Rock, Arthur	80
Roddenberry, Gene	95
Rolph, Mayor James	69
Roosevelt, President Franklin Theodore	55, 69, 70, 98
Ross, Fred Jr.	93
Royal British Air Force	73, 74
Ryan, Harris	68
Sagan, Carl	135
San Carlos Airport	134
San Francisco Bay	20
San Francisco	16, 31, 116
San Joaquin Valley	93, 115
San Jose Chamber of Commerce	70
San Jose City College	94, 98
San Jose Downtown Association	113
San Jose Evening News	38
San Jose Herald	38
San Jose Mercury Magazine	92
San Jose Mercury News	38, 92, 136
San Jose State University	22, 55, 94
San Mateo Bridge	67
Sands, Carol	131
Sanger, Hattie Emily	37
Santa Clara College	22
Santa Clara County	37, 54, 98, 115
Santa Clara University	55
Santa Lucia Mountains	17, 19
Sarnoff, David	72
Schneider, Jimmie	34
Scott, Ridley	103
Serra, Father Junípero	18, 21, 136
Shatner, William	95
Sherman, William T.	36
Shirley, Donna	78
Shockley, William	77, 80
Sierra Nevada	41
Smith, Tommie	94
Social Security	68
Soulé, Frank	17
South Bay	33, 63, 116

Soviet Union 76, 91
Sperry-Rand Company 77
St. Claire Hotel 92
St. James Park 62, 97
St. James Square 39
St. Mary's College 66
St. Matthew's Episcopal Church 65
Stabile, Judy 113
Stanford Solid State Laboratory..82
Stanford University 20, 39, 45, 60, 95
Stanford, Jane Lathrop 40-44
Stanford, Leland Jr. 41, 42
Stanford, Leland...40-44
"Star Trek" 95
Star Wars 112
Starbucks 123
Sunsweet Growers, Inc. 39
Super Bowl XVIII 103
Tan Son Nhut Airport..97
Tarter, Jill 135
Taylor, Bayard 45
Teller, Ede ....
Terman, Frederick 68
Terman, Lewis 68
Time 123
Treaty of Paris 17
Treaty of Versailles 71
Tripp, Dr. Robert Orville 33
Truman, President Harry S...75
Tupper, Gene 95
U.S. Air Force 71, 96
U.S. Army 69
U.S. Atomic Commission 75
U.S. Congress 41, 98
U.S. Embassy 97
U.S. Navy 69
U.S. Senate 40
United Farm Workers (UFW) 93
University of California, Berkeley 67
University of Santa Clara...120
USS ZRS-5 Macon 70
USS ZRS-4 Akron...70
Valentine, Don 103
Vancouver, George 21
Varga, Edina 127
Varian, Russell...73, 124
Varian, Siegurd 73
Vietnam War..95, 97
Vizcaíno, Sebastián 17
Wadell Creek 20
Walla-Walla Indians..34
Walt Disney 77, 112
Warnock, John 104, 116
Whipple, Laura 70
Willow Glen 54, 78
Winchester Mystery House 39
Winchester, Sarah 39

Winters, Bud..94
Woodside 33, 34
Works Progress Administration (WPA) 69
World War I 62, 69
World War II 55, 62, 69, 72, 98, 99, 96
Wozniak, Stephen 91, 102, 103
Ynigo Ranch 70
Ziff, Bill...129
Ziff-Davis 129
Zworykin, Vladimir..72

Partners and Web Site Index

AboveNet Communications, Inc. www.above.net 188
AITech International Corporation www.aitech.com 190
Altera Corporation www.altera.com 192
ARCOM www.arcom.com 194
Asanté Technologies, Inc. www.asante.com 196
Atmel Corporation www.atmel.com 198
ATMI www.atmi.com 200
BDO Seidman www.bdo.com 156
BEA Systems, Inc. www.beasys.com 202
Bliss Industries, Inc. www.blissindustries.com 168
Cadence Design Systems, Inc. www.cadence.com 204
Cisco Systems, Inc. www.cisco.com 206
Cornish & Carey Commercial www.ccarey.com 154
Flextronics International www.flextronics.com 208
GlobalCenter www.globalcenter.net.....................................210
Greater Bay Bancorp www.gbbk.com 158
History San José www.historysanjose.org............................176
IDT (Integrated Device Technology) www.idt.com 180
Kinetics www.kineticsgroup.com 212
Linear Technology Corporation www.linear-tech.com 184
McLarney Construction, Inc.
www.mclarneyconstruction.com 148
Palmer College of Chiropractic West www.palmer.edu 170
Peck and Hiller Company 155
PeopleSoft www.peoplesoft.com 214
San José State University www.sjsu.edu 172
San Jose Water Company www.sjwater.com 177
Valley Transportation Authority www.vta.org 174
SEMI www.semi.org 216
SideMark www.sidemark.com 150
Silicon Image www.siimage.com 218
Symmetry Corporation www.symmcorp.com 178
Technology Credit Union www.techcu.com 160
Thermal Mechanical www.thermalmech.com 152
Top Line Electronics, Corp. www.toplineelec.com 166
Trend Technologies www.trendtechnologies.com 162
University of California, Santa Cruz www.ucsc.edu................179
Webcor Builders www.webcor.com 144
Winbond www.winbond.com 220

E-TEK Dynamics www.e-tek.com.....................................Patron